Avoiding Common Pilot Errors

An Air Traffic Controller's View

Other Books in the TAB PRACTICAL FLYING SERIES

ABCs of Safe Flying—2nd Edition *by David Frazier*

The Pilot's Radio Communications Handbook—2nd Edition *by Paul E. Illman and Jay Pouzar*

The Art of Instrument Flying *by J.R. Williams*

Aircraft Systems: Understanding Your Airplane *by David A. Lombardo*

Mountain Flying *by Doug Geeting and Steve Woerner*

The Aviator's Guide to Flight Planning *by Donald J. Clausing*

The Pilot's Air Traffic Control Handbook *by Paul E. Illman*

Avoiding Common Pilot Errors

An Air Traffic Controller's View

John Stewart

TAB Books
Division of McGraw-Hill
New York San Francisco Washington, D.C. Auckland Bogotá
Caracas Lisbon London Madrid Mexico City Milan
Montreal New Delhi San Juan Singapore
Sydney Tokyo Toronto

To the memory of Irene
and the reality of Peggy.

pbk 21 22 FGR/FGR 0 9 8 7
hc 1 2 3 4 5 6 7 8 9 FGR/FGR 8 9

Library of Congress Cataloging-in-Publication Data

Stewart, John, 1945 May 3 –
Avoiding common pilot errors : an air traffic controller's view / by John Stewart.
p. cm.
Includes index.
ISBN 0-8306-1434-6 ISBN 0-8306-2434-1 (pbk.)
1. Air traffic control—United States. 2. Airplanes—Piloting.
3. Errors. I. Title.
TL725.3.T7S83 1989
629.136′6—dc20 89-31834
CIP

Edited by Carl H. Silverman

PFS
2434

Contents

Acknowledgments vii

Abbreviations vii

Introduction xi

1 Basic Pilot Errors 1

The First Error • Preflight Preparation • Making a List and Checking It Twice • Improving Communications—A Case in Point • A Difference Between Night and Day • A Tale of Two Flight Plans • Fact Not Fiction

2 Communication 17

The Right Word • Is Anybody Listening? • Phraseology • Call Signs • Pay Attention • Great Expectations • Professional Skepticism • Body Language • Flight Plan Flubs • Using and Abusing the System • Communication—Cornerstone of ATC

3 **Phraseology and Word Concepts** **41**

No Harm, No Foul • Radar Contact • Go Ahead • Urgency • Emergency • Additional Services • Clearances • Surprise, Surprise, Surprise • Very Special VFR • Cruising for a Bruising • Hurry Up and Wait • Controllers Do Care

 **ATC Equipment** **69**

Radar • Radio and Telephone Systems • Flight Data Systems

5 **Regulations and Procedures** **119**

Staying Current • Quality Control • Stop! Put Your Pencil Down • Figuring Out the FARs • Cruising Altitudes • Position Reporting • Speed Regs • TCA—Terribly Complex Airspace? • ATC Instructions vs. Pilot Authority • IFR Airfiling • Learning ATC Procedures • Ask and Ye Shall Receive • But Don't Ask for the Impossible • Saying It Right

6 **Airspace** **157**

Visibility and Cloud Clearance • Read the Regs Carefully • Airspace Anatomy • The Saga of Peggy Piper • Tale of the Phantom Control Zone • Airport Advisory Area • Military Operations • "Remain Clear of the TCA" • The Long Wait to Climb • "Remain Clear of the ARSA" • Common Sense

7 **The Future** **191**

FAA—Caught in a Crunch • Taking the Independence Avenue • Contract Employees • Radar Advances • Satellites • TCAS • MLS • Weather Technology • Sector Suites • Consolidation • Advanced Automation System • Other Developments

Final Thoughts **219**

About the Author **220**

Index **221**

Acknowledgments

I WOULD LIKE TO THANK ALL OF THE PEOPLE ON BOTH SIDES OF THE RADAR scope who helped make this project possible:

For those pilots who pulled the stunts that I recorded for posterity, I hope you recognize yourselves.

To the controllers who did likewise, I told you those things would come back to haunt you.

Special appreciation to TAB BOOKS Inc. for allowing an untried author a chance to be published.

To Mr. Ray Leader of Atlanta Tower, without whose help this book would never have come to be.

To Mr. Jeff Griffith, formerly of Atlanta Tower and now assistant air traffic manager at Tampa Tower, who taught me a lot of things, and whose knowledge and criticism of this work were invaluable.

To Mr. Clint Rogers, manager and chief pilot at Epps Air at Dekalb Peachtree Airport, for his critique and the story that you will read in Chapter 6.

To Mr. Roger Myers, of the FAA's southern region public affairs office, who reviewed this book to make sure that I got all of the i's dotted and the t's crossed.

And finally, to my wife Peggy who helped in every aspect of this book, and to my son Brad whose computer skills were invaluable.

Let's get on with the book.

Abbreviations

A/N	alphanumerics
AAS	advanced automation system
AAS	airport advisory service
ACF	area control facility
ADF	automatic direction finder
ADS	automatic dependent surveillance
AFB	Air Force base
AIM	*Airman's Information Manual*
ARSA	airport radar service area
ARTCC	air route traffic control center
ARTS	automated radar terminal (tracking) system
ASDE	airport surface detection equipment
ASR	airport surveillance radar
ATA	air traffic advisory
ATA	air traffic assistant
ATC	air traffic control
ATCS	air traffic control specialist
ATIS	automatic terminal information service
AWOS	automated weather observation system
CAF	cleared as filed
CTAF	common traffic advisory frequency
CWP	central weather processor
DCU	data control unit
DME	distance measuring equipment
EDCT	estimated departure clearance time

EFAS	enroute flight advisory service
FAR	Federal Aviation Regulation
FBO	fixed-base operator
FDEP	flight data encoding printer
FDIO	flight data input output
FRC	full route clearance
FSDO	Flight Standards District Office
FSS	flight service station
GADO	General Aviation District Office
HVOR	high VHF omnidirectional range
IBM	International Business Machines
ICSS	integrated communication switching system
IFR	instrument flight rules
ILS	instrument landing system
IR	instrument route (military)
LOM	locator outer marker
LTA	Letter To Airmen
MLS	microwave landing system
MOA	military operations area
MSL	mean sea level
MTI	moving target indicator
MVA	minimum vectoring altitude
MWP	meteorological weather processor
NAS	National Airspace System
NDB	non-directional beacon
NEXRAD	next generation radar
NOAA	National Oceanic and Atmospheric Administration
NOTAM	Notice to Airmen
NTSB	National Transportation Safety Board
NWS	National Weather Service
ODAPS	oceanic display and planning system
PAR	precision approach radar
PDR	preferred departure route
RNAV	area or inertial navigation
RWP	radar weather processor
SFA	single frequency approach
SID	standard instrument departure
STAR	standard terminal arrival route
STOL	short takeoff or landing
TCA	terminal control area
TDWR	terminal doppler weather radar
TRACON	terminal radar approach control

TRSA	terminal radar service area
TVOR	terminal VHF omnidirectional range
TWEB	transcribed weather broadcast
UCT	Coordinated Universal Time
VFR	visual flight rules
VOR	VHF omnidirectional range
VR	visual route (military)
VSCS	voice switching and control system

Introduction

THIS TEXT IS DESIGNED TO GIVE PILOTS, INSTRUCTORS, STUDENTS, AND ANY interested controllers a look at the way some pilot errors are perceived by one air traffic controller. While that is my profession and I am employed by the Federal Aviation Administration, I want to emphasize that I am writing this book as a private citizen and *not* as an FAA spokesperson. The FAA specifically disavows any connection with this or any other private enterprise publication and, except where actual FAA employees are quoted with their permission or material is obtained from other publications, all of the opinions stated in this book are mine. Now that we've gotten that out of the way we can get down to what this book is all about.

This book is about how a controller sees certain pilot actions which we would classify as pilot errors (actually those are not exactly the words that we normally use to describe those events but then this is a family-oriented book). To do this I will use events that are based on my personal experiences as a pilot, aviation instructor, and controller and supervisor in the FAA, plus over 20 years of association with the aviation business. These situations are true in the sense that similar events actually occurred; however, the pilot and airport names, and aircraft call signs, used in the text are fictional. Any resemblance of actual persons, places, or aircraft is strictly coincidental (sorry, had to get that in there, too).

To accomplish this, and for the sake of simplicity, I am going to create a fictional aviation concern and several fictional pilots that I will use in various situations throughout this book. The aviation concern will have two divisions which will be called Tandem Aviation and Tandem Airlines. Tandem Aviation will have a flight training branch for teaching pilots, and a corporate aviation branch which flies millionaires around. Most of these aircraft will have aircraft call signs that end in Tango Alpha (N123TA etc.). This is not at all unusual since all good companies are concerned with product recognition marketing strategy. Tandem Airlines will be one of the largest airlines in the world with dozens of flights leaving major airports each hour, and their company call sign is TAN(TAN1234 etc.).

Throughout the book I will introduce you to some generic pilots (are there really any other kind?) whose names just seem to fit the situation and are relevant to some of the names we use in the control tower and radar room to describe pilots. You will meet Top Gun, the macho, he-man of the airwaves and his side kick J.R., the boss's son who is sort of the Fast Eddie of aviation. We also have the general aviation (we frequently refer to them as general aggravation) triplets, Sam Cessna, Peggy Piper, and Barbara Beech. Sam is a friend of mine so he'll let me get away with this. My wife's name is Peggy, so enough said about that. Ms. Beech is named after a special lady friend of ours, named Barbara, whose nickname is similar to Beech. She would be the all-time terror of the skies if she ever got her license. I'll also introduce you to the Smart family and a few other characters that you would really have to see to believe. Just remember, this is all in fun, and yes, I do know some of the names that are used to describe air traffic controllers.

Also, in some places in this book I refer to a controller as a *he*. This is done simply for clarity, and in no way is this meant to slight the many fine female controllers with whom I have worked.

Let me begin this book by saying that, while we will be focusing on some of the less artistic things that some pilots do, this is by no means intended to be reflective of the general opinion that most controllers have of pilots, nor is it mine. We will try to poke a little fun at some of the situations in which pilots find themselves and in the process perhaps pass along some useful information that will enable you to avoid these same situations in the future. The emphasis of this book is on common pilot errors and, because of its nature, will not probe too deeply into those mistakes made by controllers. I will cover that material in my next book, titled *The One Or Two Errors Made By Controllers*. Just joking.

Pilot and controller groups are like two very good teams in the same league. They have a healthy respect for each other, a certain amount of competitiveness, and a truckload of stories about how the other team is a bunch of wimps. Like any other group of individuals who spend a lot of time in the same arena, we try not to allow the other team to take itself too seriously. Spend a little time in either camp and you will hear any group of pilots/controllers make remarks about the other group that, if taken seriously, sound like there is a genuine dislike between the two. But bring these two teams together and you have an all-star gathering that will fight shoulder-to-shoulder against the anti-aviation elements who try to cut down either group.

Throughout this book I will relate some of the "wise guy" type of remarks that we controllers make about pilots, but these comments are made tongue-in-cheek, because I know that the shoe fits on the other foot just as easily. Believe me, we have heard them all. So this book is from the "tin god in the glass house" and is written for, about, and with respect to the "prima donna zoomers in their paper airplanes."

Seriously though, the National Airspace System is a very complex mixture of routes, procedures, regulations, equipment, and equipment requirements. Inside this little world of ours there is a considerable interaction of a large number of aircraft in a small amount of airspace. This system is constantly evolving as a result of new equipment or capabilities on the part of the pilots and the aircraft, or as a result of the sometimes tragic events that occur within the system.

The aviation community is a relatively small group of individuals and, when taken in comparison to the general population as a whole, is a rather exclusive fraternity. We, as controllers, consider ourselves to be an intricate part of that fraternity, and you might even say that our very existence is dependent upon the fact that there are pilots to speak with and aircraft to control.

Most pilots don't realize that a fairly large percentage of controllers are also rated pilots. At Atlanta, for example, over one third of the controllers are also pilots, many of whom are aircraft owners and often rated in several types of large, complex aircraft. Additionally, a very large percentage of these pilot/controllers are certificated flight instructors. Even so, the average pilot knows very little about what an air traffic controller actually does. Very few pilots have actually been to an ATC facility or talked face-to-face with a controller. Most ATC facilities make a concerted effort to allow tours of the tower and radar room, and controllers frequently involve themselves in various types of pilot meetings. Despite this effort, there is still a lack of understanding between our two groups that sometimes borders on suspicion.

Some novice pilots perceive the controller to be some sort of absolute, god-like, authority figure that must be obeyed at all costs. Unfortunately, there are some controllers who believe that this should be a correct perception. But there are other pilots who seem to take considerable delight in doing as little as possible to help the controller do his/her job. There are those who will go to great lengths to "beat the system" and will do things that they themselves would consider stupid should an accident occur. Often, these pilots are simply trying to hurry and don't want to work within the system. They think they can do what they want, or get where they want to go, faster by not involving themselves in the air traffic system.

Controllers often do give instructions which do not seem advantageous to the individual pilots in terms of operating their aircraft in the fastest and most efficient manner. In the vast majority of cases though, these instructions are based on the aircraft separation requirements. Frequently other aircraft are on another frequency and their existence may be unknown to the pilot receiving the instruction. Sometimes the simple fact is that the controller is unable, for any number of reasons (usually traffic), to do what the pilot requests. In fact, one of the ways controllers cause themselves problems is by overcoordinating among themselves to accomplish pilot requests or expedite operations to hurry pilots on their way.

If this book accomplishes nothing else, I hope that it will give pilots a chance to look at the air traffic system through a different pair of eyes—eyes that see a different side of the total picture. Just remember, the system works best when we don't bend the rules and we don't cut corners. When one of us does try to use the "easy way" around established procedures, Murphy (the guy who writes the laws) always seems to be there waiting for us.

1

Basic Pilot Errors

ANY DISCUSSION OF ERRORS, WHETHER THOSE MADE BY PILOTS OR PEOPLE IN any other profession, usually revolves around three basic types of errors—those of commission, omission, and/or ignorance. The first, *commission*, is a situation where an otherwise reasonably intelligent individual does something that is known to be either wrong, dangerous, or both, for the sake of expediency pleasure, or profit. I am inclined to define this as stupidity, an action which I have always felt should be considered a capital offense. Because you are reading this book, it is obvious that you are trying to eliminate as many errors as you can, and are, as a result, an intelligent, discriminating person who doesn't fall into the first category. Therefore, any reference to errors of commission/stupidity will be made in terms of what the other guy would do, and we will simply learn from those mistakes.

The second type of error, *omission*, is typically a situation where an individual makes the type of mistake that we call ''human error.'' In this case, an individual forgets, misunderstands, doesn't hear, or in more precise terms, simply ''screws up.'' This category of error is equally as dangerous as the first but is usually correctable through proper training and the formation of good habits.

The last type of error, *ignorance*, is potentially the most dangerous because individuals will place themselves in jeopardy and not be aware of the danger. As the old saying goes, ''Not only did he not know, he didn't know that he didn't know.'' This type of error can be made by a pilot at almost any experience level, but it is frequently made by a pilot who is learning a new flying skill.

Unlike almost any other skill, learning how to fly an aircraft can create a situation with an unusual and potentially dangerous ambiguity. An individual can master the physical requirements necessary to make him/her a reasonably competent aircraft driver, and yet still know very little about what it takes to become even a marginally safe, proficient pilot.

Controllers have to contend with errors of ignorance more often than any other type, and no facility, regardless of the volume of their traffic, is immune. As an example, it is very rare that a week goes by in the air traffic control world during which a controller or flight service station specialist does not have to teach something, over the radio, to a lost pilot. This teaching can include how to operate a transponder or a piece of navigational equipment or, in some cases, how to fly the aircraft in actual instrument flying conditions. In most of these situations, the aircraft is found and directed to a safe landing, with only the pilot's pride being damaged. In others the aircraft, and sometimes the pilot, becomes a statistic. Heaven only knows how many other pilots find themselves in similar situations and do not ask for help. I can only guess that these pilots are afraid that an admission that they are in trouble will result in an appointment with that scourge of the airways, the dreaded "FAA examiner."

In most of these situations the pilot probably finds some airport and lands to find out where he (or she) is. Any hangar bum can spot these pilots a mile away. They're the ones with the blank look on their faces, looking around trying to find something, anything, with the airport or city name on it. I expect that the names of the pilots that did not get lucky and find these unknown airports can be found on the list that says, "Cause of accident unknown."

In virtually every such instance, the pilot possessed the skills necessary to pilot the aircraft—to maneuver, to land or take off, to do almost all of the physical things necessary to direct the flight of the aircraft—often with a relatively high degree of proficiency. What he did not possess however, was the understanding that there is more to piloting an aircraft than needle, ball, and airspeed. Almost any controller can tell you a dozen "war stories" about the situations mentioned above and, in most cases, the worst thing that happened to these pilots was that they, or more likely their flight instructor, received a "talking to" by someone.

In most of these cases, it is not entirely the fault of the pilot that the flight develops into an embarrassing or dangerous situation. Frequently the flight instructor fails to teach the student some very critical element of understanding. But regardless of who is at fault or who needs additional "education", I have always felt that it is better to spend an hour in the woodshed with the FAA than a somewhat longer period of time pushing up daisies.

THE FIRST ERROR

When deciding to learn to fly, people frequently make the first classical aviation error. They choose the wrong place to learn how to fly or the wrong person to teach them. Most human beings, when faced with the purchase of a product that will cost them a major portion of their annual salary, will spend a considerable amount of time and energy investigating this purchase from almost every conceivable angle. Unfortunately, when it comes to learning to fly, an otherwise intelligent person will give thousands of dollars to someone he or she knows nothing about, trust their judgement, and follow their teachings and examples. They usually do this without having the slightest notion of whether or not they are receiving a quality product for their money.

Why? For those who have never flown, there is a mystique about flying. To the uninitiated, often naive student, *certificated flight instructor* means that the individual who possesses this rating must know everything worth knowing about flying and what he or she teaches you is all that you need to learn. For those of you who are learning to fly, contemplating learning to fly, working on advanced ratings, or who think that you have been taught all there is to know about flying, be advised: this is not necessarily the case.

I'm not knocking flight instructors. I hold those ratings myself and know, or have known, several outstanding instructors over my aviation career. Unfortunately, I also know some who fall a little short of the outstanding category. Rather, I'm knocking the system itself and possibly the way we are teaching everything. Ours is a "fast everything" society, dedicated to doing everything quickly and learning just enough to be able to, as they say, "pass the course."

The pressure to teach the student quickly is very real. It is dictated by the simple economic fact that students will, in most cases, learn where the cost is lower and where they can realize their goal, of learning how to fly, faster. Most people want to take their flight training at the same time that they take the ground school portion of their training. As a result, they frequently learn the mechanics of how to do something before they understand why they are able to do it. Complicating this process is the availability of quick study courses that promise to teach you, in three days or less, all you need to learn in order to pass the written test for your license. A three-day course directed at teaching the written exam is an excellent *refresher course* for someone who has studied and understands the material yet needs some practical test-taking experience and a last-minute review. Most of these courses teach exactly what they promise to teach, a preparation for the written exam. Some do not even make an effort to do that and few, if any, achieve the level of learning necessary to build a fundamental understanding of all of the material contained in their ground school course.

Because of the fact that young pilots often use a job as a flight instructor to build flying hours toward an eventual career with the airlines, many students are taught how to fly by instructors who have only a few hundred hours of flying experience themselves. None of these factors are necessarily bad in and of themselves, although they do have some major drawbacks.

The selection of an instructor is important, but don't begin with the idea that older is better. Select your instructor based on what you are able to learn about the abilities of those available. Learning to fly from a young, energetic flight instructor can be a rewarding experience. When associated with a quality flight-training program that monitors the performance of all instructors, provides them with continued learning experiences, and uses a system of check rides by qualified examiners to ensure that what is being taught is correct and of high quality, these instructors are frequently among the best around. In an ideal situation, both the student and the flight instructor should have learned something at the end of each lesson. It is not at all improper for a teacher to enhance his or her career while teaching.

While it may seem to be a somewhat unfair concept, you, the student, regardless of the experience level that you have, bear the responsibility for selecting a quality flight-training program. You should treat this process the same as you would any other major purchase. Shop around, compare what is being offered for the money, and talk to some of the people around the airport about the reputation of the flight school and its employees. Ask for a list of names of people who have taken flight instruction at the school and then talk to them about the quality of their training.

Check out the reputation of the flight-training facility with the FAA offices in your area and check with the Better Business Bureau. Ask to see the credentials of the staff, and find out how long the senior staff has been with this organization. You could also check them out with the competition but, unless bad comments can be substantiated with facts, I would only listen to the good things that the competition has to say.

Once you have chosen your flight school, don't drop your diligence. If your instructor frequently cancels your lesson in order to fly a charter, it is time to ask for another instructor. If this tendency prevails throughout the staff, a new flight school is in order. Your instructor should be given the latitude of choosing what is to be learned, how soon he or she is willing to allow you to do things on your own, and for the most part, you should trust your instructor's judgement as to the learning pace. If your instruction still seems to be disjointed and there is no reasonable explanation as to why, or if your instructor does not seem to be able to remember where you are in your training progress, perhaps you should listen to your instincts to leave. At the very least, seek out the advice of someone you trust.

PREFLIGHT PREPARATION

Another common error that pilots of all levels make is getting into the aircraft unprepared for what they are about to do. They are eager to begin the flight or, in a training environment, the flight portion of their training. But frequently they do not take the necessary time to learn the background information associated with the route that they will fly or the basic knowledge required for the new skill that they hope to learn. When applied to routine cross-country or business flights, this means that the pilots will be spending a large portion of the flight time, when they should be watching their flying, working on the things that should have been done on the ground. In a training environment, this lack of planning forces them to try to learn the theory behind a particular maneuver at the same time that they are developing the physical skills required to fly the aircraft through that maneuver. In either case and regardless of whose "fault" it is, the result is that more aircraft find their way into the system, each piloted by at least one person who is not keeping up with the things that are essential to ensure a safe flight.

What most pilots do not realize is that the fact that they are "behind the aircraft" comes through loud and clear over the radio. Every controller is familiar with the sequence of events that begins when the pilot's voice starts to become strained. It continues as the pilot begins to miss radio calls or replys in a manner that leaves the controller wondering if controller and pilot are talking about the same thing. Hopefully, it concludes when a confident (instructor) voice takes over the radio duties from the totally befuddled pilot. While this sequence of events is amusing in retrospect, the controller is left wondering about the safety of the flight until the instructor's voice comes over the radio. Sometimes the insecurity even remains in the controller's mind after that point.

Frequently, the net result of this concern will be that the controller will give this aircraft kid-glove handling and not try to rush the pilot into a particular maneuver. While this is a positive action in terms of keeping the pilot more relaxed and giving him/her more time and more room to complete the maneuver, it does have some negative aspects. First, the pilot gains a somewhat false sense of security that the controller will always look out for him. Second, the length of time that a pilot needs to go from point to point is going to increase. (This is particularly relevant in a training environment because the total number of maneuvers that can be performed during a given time period is reduced because the flight path described for each maneuver is going to be longer and more time consuming.)

MAKING A LIST AND CHECKING IT TWICE

One of the first things that a controller who has been assigned to a new facility learns is the registration numbers of the aircraft in the area that are used for flight training. We also develop a mental list of those aircraft that everyone has placed on their "watch out for this guy" list. This action is not something that is specifically taught, rather it is something that each controller has learned to look for in planning control actions.

Controllers know that these aircraft are frequently piloted by inexperienced pilots or student pilots of varying skill levels, and that when they are working these aircraft they should take a look at how several maneuvers are performed before they attempt to place that aircraft into a tight situation. Controllers have to plan a sequence for departures and/or traffic pattern arrivals or approaches that is several minutes ahead of where the aircraft are when the planning begins. At this point, a turn of 5 or 10 degrees given to aircraft A is often sufficient to build a hole big enough for aircraft B to conduct another maneuver. This maneuver could be another touch and go or a complete instrument approach. Without this advance planning, aircraft B would have to fly the extra five miles to get behind aircraft A.

When the controller does this, he or she must be reasonably confident that aircraft B will be able to execute the maneuver in a normal manner. If the aircraft do not do as expected, the controller will shortly find himself or herself in a position of having to break one or both of the aircraft out to prevent an incident or accident. When this happens, the controller will be reluctant to try it again with that same aircraft. This type of planning only works when the controller knows what to expect, or is at least prepared for the adverse possibility because they know the aircraft—hence the reason for learning the local aircraft.

When pilots fly their aircraft to another airport, controlled by a different approach control or air route traffic control center (ARTCC) facility, they tend to lose this home court advantage. The controllers at the other facility, just like the ones at the home base, have to treat each transient aircraft as though it were piloted by a professional pilot unless there is a reason to do otherwise. In this situation, the controller will set up the sequences to give the most expeditious handling to the most aircraft over a given period of time. By the time the controllers at the other facility start to hear the telltale signs of inexperience or confusion, they are often committed to a so-called "airshow." This is a situation where aircraft are scattered away from one another to ensure separation. One way to reduce the potential for this type of situation is through education.

IMPROVING COMMUNICATIONS—A CASE IN POINT

During my time as an ATC training specialist, my facility spent a considerable amount of time and energy working with flight schools, flight instructors, and other frequent users of our services to develop a series of techniques that would reduce the delays and develop a sense of confidence between controllers and pilots. The procedures were all voluntary and centered on the concept that the more information we knew about each other, the more likely we were to work together as a team. After these meetings, instructor pilots began to file instrument flight rules (IFR) flight plans that initiated at some point in our facility's airspace near the designated practice areas rather than at the departure-point airport. The idea behind this was to allow the pilot and instructor adequate time to practice maneuvers in visual flight rules (VFR) flight conditions before they started back to the facility for practice IFR approaches. This was important because most approach control facilities simply do not have enough airspace to set aside a large area for use in practice maneuvers while providing IFR separation. If we did allow this type of activity, we would have to protect an area three to five miles around an aircraft that is maneuvering all over the sky at varying altitudes. Try to picture in your mind how much airspace just one of these aircraft would render unusable for other operations.

There were some additional benefits gained by advance filing as well. First, the pilot was sure that *all* of the information regarding that flight was recorded with the FSS.

When a pilot airfiles with a controller, that controller inputs only the minimum information he needs (The rest is relegated to the ATC facility's audio tape of the pilot/controller conversation.) I will talk more about this later. Second, the pilot avoided the possibility that the controller would be too busy to copy the flight plan on the frequency.

One of the things that we agreed should be included with the flight plan was information, contained in the remarks section, that identified the type operation being conducted. For example, when the flight plan called for practice approaches with an instructor on board, the abbreviation "Dual Prac App" was included in that section. This information informed the controller that an instructor was in the aircraft and tended to give the controller the confidence that the aircraft could be expected to perform critical maneuvers when needed.

Once the controllers began to recognize the instructors' voices, and instructors began to see the advantages of providing the controllers with relevant information, a whole range of additional services began to develop. Instructors would call the appropriate controller and tell him/her that they were on their way to the practice area and would want to pick up their IFR flight plan at a specific time. This allowed the controller a chance to do some advance planning and frequently, to identify the aircraft at that point. The controller would start a track on the automated radar terminal (tracking) system (ARTS) and provide the flight with VFR flight following and traffic advisories during the maneuver phase of the flight.

Controllers would also inform the pilots of what transponder code they could expect to be assigned for the IFR portion of the flight and where it would be most advantageous to be when they were ready for the practice approaches. Occasionally, the instructor would decide that the student needed more work on the maneuvers than had been anticipated and that they would not conduct the practice approaches. He or she would then notify the approach controller to cancel the flight plan, thus allowing more room in the computer for actual traffic. This program created a lot of good public relations between the pilots and controllers and, except for an occasional thorn in the roses, generally reduced the delays, and increased the number of training maneuvers accomplished in a typical training flight.

Similar procedures were developed for use in the traffic pattern training environment. Instructors were encouraged to notify the tower that an instructor was on board the aircraft. In these situations, controllers generally knew what to expect from the pilots in terms of response time, and the increased productivity gave instructors confidence in the program. They even began to use the controller as a training tool. They would ask the controller to give the student some unusual maneuvers during the training session so that he or she could gain experience and learn to be prepared for any eventuality at a different airport. These maneuvers ranged from 360-degree turns on the downwind, nonstandard traffic patterns, short or extended patterns, changes in runway assignments, and other unexpected actions up to and including a last-second go-around.

Individual aircraft owners or operators who were familiar with the program also benefited from the increased awareness on the part of the local pilots and controllers and began to develop techniques that passed along useful information to the controller. Pilots began to use phrases such as "locally based" or "we can accept a short approach" and began to learn the voices and operating initials of some of the controllers. (Each controller has a set of two initials which they use to identify themselves on every tape-recorded telephone communication.) They would say something like, "Cessna 123TA, at the general aviation ramp ready to taxi, how's it going J.S.?" This use of an informality on the initial call made the controller aware that they were familiar with the area. Pilots also began to use VFR reporting points religiously and spent several minutes listening to the frequency before they made the initial call to the controller. Everyone benefited from this increased awareness because controllers are more likely to rely on pilots' abilities when they know what those abilities are. As I explained earlier, this knowledge can dictate the sequence or determine the size of the arrival or departure intervals needed to accommodate those aircraft.

Most FAA facilities have some type of formal or informal program similar to the one mentioned above. Controllers also participate in pilot/controller forums and other educational programs sponsored by aviation groups. Unfortunately, most of the pilot participants at these events are the same people every time, and they are rarely the individuals who cause the problems.

Among the surprising things a pilot learns when he or she visits an ATC facility are the limitations of the equipment and information resources that controllers have at their disposal. Pilots seem to be surprised that we do not always possess the types of weather radar that can be found on the local TV news, that we do not have instant access to all of the information about a particular flight, or that we are unable to respond to all of the needs of the pilot.

Frequently, the types of questions that pilots ask or the information requests that they make only serve to underscore this fact and waste both the pilots' and controllers' time. This situation causes the controller to wonder what kind of instruction these pilots have received. When placed in these situations, the controller is often faced with having to choose between one of two unattractive options. He or she must either do a considerable amount of work to provide these pilots with the information that they require or risk a public relations failure by instructing the pilot to contact the appropriate authority and obtain the information on their own.

Most of the pilot requests seem to be relatively simple, and the pilots are frustrated and often angry when controllers appear unwilling to comply with their requests or tell them to contact a flight service station (FSS) or other agency. Most controllers understand that the pilot is very busy flying an aircraft during this exchange and that he or she is just trying to obtain information or services as easily as possible. We also know that many of the requests that we receive are the result of improper planning on the part of the pilot or a lack of understanding of how the system works.

Let's take two simple situations involving pilot requests, analyze the circumstances of the requests, and determine what the pilot or controller has to do to accomplish the requests. We will select worst-case situations typical of the the requests that controllers receive from VFR pilots and from an IFR pilot. (We won't even talk about the pilot who wants you to call his wife and tell her that he will be late for dinner.)

A DIFFERENCE BETWEEN NIGHT AND DAY

Semi-experienced pilot, U.R. Smart, arrives home from work late Friday afternoon and makes a snap decision to go to a weekend vacation resort near a very large city to celebrate the big promotion that he was just given. After loading the spouse, the three kids, and the St. Bernard into their four-place aircraft and departing without checking anything, including the weather, their flight proceeds uneventfully to the destination. The controllers are very helpful to this pilot, offering radar vectors when the aircraft drifts off course, handing the aircraft off to the next controller enroute, and even providing the frequency of the new VOR that has replaced the one on his two-year-old chart. When the Smarts arrive near their destination, it is almost midnight. They are cleared through the terminal control area (TCA) are given helpful radar vectors direct from the last VOR to their destination airport near the Grandest Resort in the World.

When Smart loads the family back into the aircraft on Monday morning, the spouse is exhausted, the children are crying from severe sunburn, and the dog is sick. Unknown to our favorite airplane driver, almost everything else has changed as well. Undaunted, our denizen of the air turns on the radio, which is still tuned to the last frequency that was used on their way to the resort, and heads out in the opposite direction, again without checking anything.

After waiting for a break on the now very busy frequency, U.R. recites all the information that any controller could possibly ever want to know. The reply from the controller is a terse "Remain clear of the Very Large City TCA and contact approach on frequency 119.9." Smart is not really sure what the controller wants, but decides that discretion is the better part of valor and starts looking at the map while trying to tune in the new frequency. As U.R. is doing this, the spouse is pointing at the L-1011 that is now visible through almost every window in the aircraft. (The kids are still crying and the dog is still sick.)

By the time Smart finally contacts the appropriate controller and requests vectors direct to the home airport, the controllers are only interested in getting this aircraft out of the way of the dozens of other planes. They will probably not trust this pilot to do anything else correctly, and they are going to be very busy coordinating the location of this aircraft with the controllers in whose airspace the aircraft is operating. The probable result is that the aircraft will be vectored around the TCA at the cost of many additional flying miles. In this circumstance, the controllers will probably be too busy to accomplish the additional coordination required to move the aircraft through the TCA. Additionally, at this point in their traffic situation, they are probably still trying to recover from the ''airshow'' caused by Smart's aircraft in the first place. If the pilot penetrated the TCA without authorization, he will probably be asked to contact the facility to explain the situation after he lands. Based on that conversation, the information may be turned over to the Flight Standards District Office (FSDO), and they will investigate the situation and possibly take punitive action against the pilot. Don't get the idea that you can get away with these actions by simply not calling. We will turn the data over to FSDO anyway, and they will still talk to you. In this situation, it is not likely that they will be as friendly as they normally are.

Our hypothetical pilot is going to remember this situation angrily, and firmly believe that the controller was being arbitrary in forcing the aircraft to divert around the TCA. After all, he did exactly the same thing on Friday. But what he doesn't realize is that the two events he experienced were totally different. When the aircraft arrived in the TCA at midnight on Friday, most of the traffic was gone, the existing traffic was arriving from a direction that did not conflict with Smart's flight path, and several radar positions were combined on one frequency. On Monday morning, when the Smarts took off, the Very Large City airport was in the middle of a very busy arrival/departure rush, they were landing from a direction that conflicted with Smart's flight path, and all available radar displays were in use.

This last circumstance frequently means that pilots operating in the same airspace will have to talk to a different controller on a different frequency at different times of the day. As the traffic workload increases at a facility, airspace is split into smaller segments, each controlled by a different controller on a different frequency. The only limiting factors in this procedure are the traffic requirements and the number of positions/controllers available. So just because you used a particular frequency at midnight on Friday, does not necessarily mean that the same airspace is controlled by the same controller at 9:00 A.M. on Monday.

This is a classic example of how an unfortunate situation can develop into a case of one pilot becoming soured on the ATC system. Although the scenario seems extreme, believe me, it is something that happens all too often. Had the Smarts filed IFR or asked for radar advisories while still on the ground at the small airport, they would have been given the correct frequencies, clearances for departing that airport and entering the TCA, and most importantly, the next controller would have known that they were coming. While the FAA hires the best people it can get for all of its positions, only the secretaries are required to be telepathic.

These concepts of advance planning, proper usage of equipment, and once again, a good understanding of how the system works, are even more important to the IFR pilot. Perhaps the most frustrating and annoying thing that pilots do to controllers in the IFR realm has more to do with the timing and correctness of their requests, the accuracy of the information provided to the controller, and the proper use of their equipment, than with what they want.

Now, let's take a look at how circumstances can combine to turn a relatively routine flight into a situation where controllers are willing to start a pilot assassination fund.

A TALE OF TWO FLIGHT PLANS

Ashley Conscientious, a corporate pilot for Tandem Aviation (the corporate flight department of Tandem Widgets International) arrives at the airport several hours before his scheduled flight from his company's regional office to its headquarters in Georgia. He checks the weather, files the appropriate flight plan for a BE90 King Air N50TA, requesting 10,000 feet to Satellite Airport, Georgia. He then gives the aircraft a thorough preflight, contacts the tower on the clearance delivery frequency, and picks up his clearance. As he goes back into the pilot lounge he notices that another company aircraft, N60TA, the Super King Air BE200, is sitting on the ramp being refueled and restocked.

In the lounge he notices that Smiley Incompetent, the newest company pilot, is on the phone. Smiley, who likes to be called by the nickname "Top Gun," gives him a thumbs-up sign and goes back to talking on the phone. Just then, the Big Boss and several others Ashley had flown to the regional office come rushing into the lounge.

"Let's go, Ashley," says B.B."We gotta get back home before the market closes."

As they rush out to the ramp, Ashley realizes that they are heading for N60TA and hurries to catch up to B.B.

After he explains that he has filed a flight plan for the BE90, the Big Boss says, "I had Smiley take care of all that flight plan change routine, the paperwork is in the airplane, and he did all of that pre-trip stuff you guys think is so important. You can pick up your bags from him when we get home, now let's get cracking."

As Ashley slips into the left seat he notices that his copilot is J.R., the Boss's son, who fancies himself as the only pilot with the "Right Stuff." J.R. hands him a copy of the flight plan and indicates that all of the preflight operations have been completed.

Ashley starts the engines and begins a call to the tower for taxi clearance. Just as he starts the transmission, B.B. pokes his head into the cockpit and asks a question to which Ashley nods his head yes several times. The resultant transmission sounds something like this. ''Airport ground, this ......xty Tango Alpha, general aviation ramp, IFR, ready for taxi.''

What Ashley doesn't realize is that Smiley improperly used local time when filing his flight plan, instead of Coordinated Universal Time (UTC). The FSS specialists, of course, assumed the time figures were UTC. They know that corporate pilots frequently file a flight plan for the next day shortly after landing. Of course the fact that they gave Smiley a weather briefing for tomorrow's weather went right over his head. Smiley also did not cancel Ashley's flight plan because he intended to use that one for himself. He didn't think that it was important to change the flight plan so that the correct pilot's name was included. After all, he didn't plan on having an accident. This means that the flight plan for 60TA aircraft will not be in the computer until twenty hours later and the old flight plan for N50TA, already read by the clearance delivery controller, is still the only one sitting in front of the ground controller.

The ground controller, who is used to garbled transmissions, looks through her binoculars at the GA ramp over a mile and a half away and sees what looks like a T-Tailed King Air moving on the ramp. Looking into her flight progress strip bay, she finds a flight plan for a 50TA, obviously one of the few T-Tailed BE90s, and sees that this is the only flight plan that has been read by clearance delivery. She replies ''November Five Zero Tango Alpha taxi to Runway Two Zero Left,'' and since the pilot did not indicate that he had the current automatic terminal information service (ATIS) code, she also reads the appropriate weather information.

Unfortunately, Ashley is busy trying to sort out the mess in Smiley's flight bag while trying to locate the airport diagram and misses the fact that she said N50TA. J.R. is blissfully ignorant of just about everything except those things that make life easier, so he shortens the call sign to Zero Tango Alpha (the *AIM* says you should only do this *after* the controller has abbreviated the call sign) and acknowledges the transmission. Ashley begins to taxi the aircraft to the runway and to do the pre-takeoff checks without checklists, which have disappeared.

At the proper time, the tower will assign a departure heading and clear N0TA (controllers will call you what you call yourself unless it will result in confusion) for takeoff and instruct the pilot to contact the departure controller. Ashley guides the aircraft into the air without spilling the martinis that are flowing in the back, and just after he retracts the gear, the tower advises him to contact departure. As he starts the turn, he realizes that this is not a normal departure procedure for this type aircraft, but before he can say something to the tower, J.R. switches to departure, and problem number two pops up.

The departure controller asks the pilots what four-digit code they have selected in their transponder. J.R. reads the code (the same one that this aircraft had been assigned on the way in) to the controller who then gives them a totally different code to squawk. This new code sounds vaguely familiar to Ashley, but he has flown this route so many times that everything sounds familiar. Ashley is just about to question the heading when the controller assigns them a new heading that is in the general direction they want to go and authorizes a climb to 10,000 feet. Ashley dismisses the original problem as probably a rookie controller in the tower and continues on with the flight.

With J.R. doing the actual flying, Ashley now takes the time to really study the flight plan and realizes that B.B. has scheduled a short stopover in Jacksonville, Florida, which is IFR. He notes with some chagrin that Smiley has forgotten to bring the Florida charts in his chart case. Ashley has been in the air for almost 10 minutes now at a reduced airspeed of 250 knots and he is already annoyed at what has happened so far. He is just about to get angry at the controller for delaying their climb to their filed altitude of 23,000 when the controller advises them to change frequencies to the next terminal radar approach control facility.

Ashley grabs the mike and demands to know why they haven't been cleared to a higher altitude and why they are being sent tower enroute instead of being handed off to the ARTCC. The controller responds that the information at his disposal shows the requested altitude is 10,000 and the filed route is appropriate for tower enroute navigation to Satellite Airport, Georgia. Suddenly the light dawns on Ashley, the familiar transponder code is the one he was given for N50TA, the departure turn he received is the one assigned to low-performance aircraft such as the BE90, and the route he is flying is on a flight plan to the wrong airport at the wrong altitude.

Impossible for this situation to happen? Only the names and the call signs have been changed to protect the guilty.

Behind the Scenes

Now let's take a look at what kinds of havoc this scenario caused in the air traffic control system.

Smiley is going to call for his clearance only to find that one of his company aircraft has already used the flight plan and he will have to refile and wait for the computer to process the information. Knowing Smiley, he will ask a busy clearance delivery controller to do it for him and tie up a frequency for several minutes while the controller copies the data. If the controller is nice enough (crazy enough) to do this, it will take several more minutes of his/her time to type the flight plan into the computer while other aircraft have to wait before receiving their clearance. Smiley's original flight plan was activated when the controller assigned that transponder code to N60TA thinking that he was talking to N50TA.

The ARTS system activates an IFR flight plan when it finds an aircraft broadcasting a transponder code that matches a legitimate flight plan in its storage area. It then sends that departure message to all of the facilities enroute that will talk to this aircraft and prepares them to expect this aircraft at a certain time based on the filed speed of the aircraft. Once started, the only way to stop this action is to put the flight plan on indefinite hold or remove it from the system.

Twenty hours from now, the computer will generate a flight plan for N60TA that will waste controller and computer time for at least two hours. This is the normal length of time that a flight plan stays in suspense before it's automatically dropped. When Ashley took off with the wrong transponder code selected, there is a strong probability that this action activated another flight plan for a totally different aircraft. This brings to three, the total number of flight plans that someone in the air traffic system is going to have to refile.

The greatest workload, however, comes from the fact that this aircraft is boring holes in the sky at 250 knots, very close to the edge of the controller's airspace and radar coverage area. These controllers are forced to keep the aircraft in their airspace while they advise the surrounding facilities of the situation. This involves coordinating with the ARTCC to verbally file a limited flight plan which will allow the King Air to continue on its way to Florida. They must also advise the next terminal facility that the aircraft is not coming and remove the original flight plan from the computer. They must also do this while controlling all of the other aircraft that they had when this mess began.

During this entire process, the pilot is exerting pressure on the controllers to hurry because B.B. is breathing down his neck. If this isn't enough, remember that we have not even begun to deal with the IFR approach into Jacksonville without a copy of the approach plate.

FACT NOT FICTION

While these two scenarios might seem to be fictional comedies of errors, they are typical of the problems that controllers face every day. I have also tried to fill in the background a bit to demonstrate how easy it is to allow normal everyday pressures to add up and magnify small, individual errors. Every one of these errors points out a breakdown in some aspect of the basic aviation skills that each pilot should possess and exercise prior to each flight. These errors are very common and they contribute dramatically to the workload on the ATC system.

At an airport like Atlanta, where we handle as many as 3000 operations a day, it is not unusual for 10 to 20 aircraft to take off with the wrong transponder code selected. While this seems like a small item, each instance results in two to three additional transmissions, at least one additional telephone communication to the ARTCC sector that receives a departure message, and some work to restore the lost flight plan.

In the last example, I gave you a glimpse of what has become another controller nightmare, similar-sounding call signs. Large corporations like to have their aircraft named after their company. Like Tandem Widgets mentioned earlier, it is not unusual for a company to have several aircraft with call signs that end with the same two alphabet characters and similar numeric characters. Frequently, these aircraft are in the air at the same time talking to the same controller and a transmission to N50TA will result in the question, ''was that for 60TA'' as a reply. We will cover more of this in the next chapter, but as the old saying goes, ''If I had a dollar for every time this happened, I wouldn't need this job.''

No pilot has immunity in this situation. Flight instructors may use the same two or three aircraft several times a day, and airline pilots frequently use the same call sign for more than one segment of their trip and then have to change to an entirely different sequence of numbers. It takes some mental alertness to spend several hours and/or thousands of miles responding to an airline call sign ''Tandem 279'' then ignore control instructions to that flight because you are now flying ''Tandem 543.'' If equipment capabilities ever reach the level necessary, I think each pilot should be assigned an N number and a transponder code when they get their license, and carry those numbers to their grave before they are assigned to anyone else.

The bottom line in aviation is ''know, understand, and comply with the basics.'' Learn as much as you can about the the skills needed to be a good pilot before you climb into an aircraft. Learn enough about meteorology to be able to predict the type of weather you may encounter, or at least know how to obtain weather information. Learn the ATC system well enough to understand the pitfalls and potential errors. Use checklists religiously so that you don't make the simple mistakes that add up to major problems.

Once you climb into the aircraft, remember that you have the responsibility to be the best student, commercial, instructor, or airline pilot that you can be. Human error is the one thing that we are unable to build or legislate out of the system, so leave that ''kick the tires, light the fires, top gun'' mentality in the pilot lounge, and take the professional pilot baggage into the aircraft with you.

1523

2

Communication

WEBSTER DEFINES COMMUNICATION AS AN ''EXCHANGE OF INFORMATION, ideas, or knowledge.'' This definition tends to take for granted, or at least imply, the concept of understanding, which is, of course, the most important aspect of the information transaction. When we say that ''we failed to communicate,'' we are really saying that even though we exchanged information, the exchange did not result in understanding. It could also mean that the participants in the exchange came away with a different ''understanding'' of the same information.

In the ATC system, communication usually means some form of radio transmission exchange between pilot(s) and controller(s) where the participants never actually see each other. This is unfortunate because most people have been conditioned to use such things as gestures or body language, inflections, and eye contact to fill in the grey areas associated with the spoken word. This use of nonverbal enhancers enforces and clarifies the intent of the communication. It is important to point out that these actions need not be limited to the enhancement of the spoken word and, when necessary, can stand alone as a type of communication. Sometimes a smile and a shrug can say volumes about what we mean.

THE RIGHT WORD

We will deal with several types of communication techniques in this chapter, but let us begin by discussing the correct use of the spoken word. We frequently use words that have more than one meaning in the English language and use them in a context that does not specifically define which meaning is being used. Let me give you an example of how this ambiguity plus a little inattentiveness can work against you in an aviation environment.

Visualize, if you will, a picture of a training aircraft (N123TA) which is in the last stages of a practice instrument landing system (ILS) approach and is on short final to Runway 18 Left at Medianville airport. The control tower has been informed that the pilot wishes to conduct another practice approach and that he or she would like to execute the full instrument approach procedure. Traffic permitting, the most efficient way for the pilots to do this procedure is to execute a missed approach which allows them to proceed directly back to the locator outer marker (LOM) and begin the procedure turn portion of the published approach from that point. Since controllers know this, they will evaluate their traffic picture and construct a control instruction based on the wishes of the pilot and the realities of the current traffic. Having already done this, our hypothetical controller issues the following instructions when the aircraft reports missed approach:

> N123TA turn left heading 360, proceed direct to the Medianville LOM, cross Medianville outbound at 3000 feet, cleared for an ILS approach to Runway 18 Left.

This is not really a complicated instruction but, when a student pilot is operating an aircraft under the hood or using some other type of IFR simulation, their concentration is sometimes so intense that they may miss a key element in a transmission. In this case, our hapless student is not absolutely sure that he heard the direction of turn so he asks the instructor "did she say left turn to 360?"

At this point in the flight the instructor is busily writing down all of the mistakes that the student made on his last approach so that they can be covered in the postflight briefing. He responds, "Right" (meaning "that's correct").

Now, to complete the picture, also visualize a very large aircraft just beginning to reach takeoff speed on Runway 18 Right. This is what I mean when I say that Murphy will get you. Had the instructor said "that's correct" instead of "right," the potential for misunderstanding would have been eliminated and the student pilot could not have gone in the wrong, or "right," direction. This situation is further complicated by the fact that the student asked the instructor for clarification instead of asking the controller because the student did not want to appear inattentive or foolish.

IS ANYBODY LISTENING?

If you have spent any time in an aircraft, you have probably seen the situation where a control instruction is issued over the frequency and the pilot looks around at the other occupants and says, "Was that for us?" Most of the other occupants were not paying any more attention than the pilot so the response is usually a shrugged shoulder and a blank look. Pilots will then either wait for the controller to repeat the instruction, "fess up" to not paying attention, or use some ploy to give the impression that it wasn't their fault that the instruction was missed.

This last action is becoming an increasingly popular and sometimes valid excuse. At a busy facility with a lot of frequency congestion, two individuals will sometimes transmit at the same time and the third party, for whom the instruction was intended, will only hear a loud squeal over the frequency. The usual pilot response to this situation is to say something like, "you were blocked," and the controller will repeat the instruction.

The controller is listening to the frequency using a headset and, unless the other person begins and ends their transmission at exactly the same time as the controller (very unlikely), it will be immediately obvious that part or all of the controller's transmission was missed and he or she will repeat the instruction. Most controllers will tell you that they have heard this "you were blocked" transmission hundreds of times and occasionally from the only aircraft that they had on the frequency.

In most ATC environments, the controller issues control instructions based on a timing requirement that will place an aircraft on a flight path that will result in the most efficient use of airspace. As an example, a control instruction for a turn to the final approach course is calculated so that the aircraft will be established on final at the ideal distance behind its traffic. The 5- to 8-second delay caused by an inattentive pilot will destroy this sequence and result either in a delay for every aircraft that follows or a scramble on the part of the controller to change the sequence. Two or three such misses will find the controller privately commenting on the marital status of the pilot's parents.

I have been in the cockpit of some very sophisticated aircraft operating near some of the busiest airports in the world and have been amazed at the lack of attention that is paid to the traffic environment during these flights. Pilots should use headsets instead of overhead speakers throughout the entire flight or at least while in a very busy terminal environment. Even though most of the transmissions are directed at other aircraft, pilots should not consider this a distraction to their discussions on the latest stock market prices.

They should use this information to develop an awareness of where they are in the traffic sequence. This is what we call "getting the picture." A sharp pilot can inform the controller of this awareness in a lot of subtle ways and this awareness almost always results in a savings to the pilot in terms of time or fuel or both. As I have indicated before, if the controller is reasonably confident that a pilot will perform a critical maneuver at the proper time, that controller is more likely to build a hole and alter the sequence to accommodate this aircraft.

I don't want anyone to get the idea that this technique is any type of reward for paying attention, because it is not. The ability to "shoot the gap" as we call it, is a conservation of time and airspace for the controller and a technique that significantly reduces the workload on all concerned. However, we cannot use this technique unless we can be assured that it will work, so the pilots who pay attention and are able to communicate that attentiveness will be the ones that we use to fill the gaps. The lesson here is that communication is at least 50 percent listening and watching. Now let's discuss the other 50 percent.

PHRASEOLOGY

When we are talking, the emphasis should be on saying the correct thing at the proper time so that both pilot and controller have the same understanding. The FAA publishes several documents which deal specifically with proper phraseology for pilots and controllers, but the most important of these is the Pilot/Controller Glossary of the *Airman's Information Manual* (*AIM*). If every pilot and controller would memorize this section of the *AIM* and use the words found in it exactly as they were intended to be used, I could eliminate this chapter and this book would be slightly less expensive. But, alas, we must all pay for our mistakes and someone has to profit from those errors, so on we go.

Roger, Wilco, Over, and Out

Over the years, pilots and controllers have developed a phraseology of their own and have attached meanings to words that only exist in the world of aviation. Unfortunately, pilots have also allowed habits to develop that use some of those phrases or words inappropriately, while other aviation phrases, which mean exactly what we want to say, have fallen into disuse. One of the most frequently used words in the aviation language, and the one most frequently misused, is the word "Roger." By definition in the *AIM*, this word means "I have received all of your last transmission." The *AIM* goes on to say that the word should not be used to answer a question which requires a yes or no answer. "Roger" in the past was defined as meaning "I have received and understood your last transmission," but the definition was changed to emphasize the fact that it was not an answer. (Perhaps because the "understanding" part was often clearly missing.) In most cases, a pilot response of "Roger" to a question requires that the question be repeated. This wastes time. A simple example of this is the following exchange:

Controller: Cessna 345TA, do you have the airport in sight?

Pilot: *Roger. (Some pilots use the technique of replying only with their call sign, e.g. "Cessna 5TA.")*

In this situation the controller cannot take the next logical step and tell the aircraft to proceed to the airport. The controller also cannot clear the aircraft for a visual approach because he or she still does not know whether the pilot has sighted the airport. Generally pilots mean yes when they use their call sign or "Roger" in this manner, but controllers learned long ago not to bet their careers on what they think pilots might mean.

In addition to the problem of wasted time, there are situations where the use of "Roger" as an answer is potentially dangerous. One of the most critical of these operations is the runway crossing situation. When controllers do not want a pilot to cross an active runway while they are taxiing on the airport, they are required to issue an instruction to hold short of that runway. If the pilot uses "Roger" as an acknowledgment of that instruction, nothing has been agreed upon between pilot and controller.

Frequently, the controller is predicating separation upon the pilot who agrees to hold short of a runway on which another aircraft is departing. When the response is "Roger," the controller is faced with having to repeat the instruction while the second aircraft is delayed. It is in this type of situation where another aviation term, which seems to have fallen out of favor, could be used to great advantage.

The term "Wilco" means "I have received your message, understand it, and will comply." This simple phrase serves as a complete reply and an agreement on the part of the pilot to do what the controller has asked. Most pilots seem to shy away from the use of this term (perhaps they are afraid to sound too much like Buck Rogers) or they do not know what it means.

The critical point that needs to be made is that you should use the correct word in the correct circumstance. When you want to say yes, use the word "affirmative." When you want to say no, use the word "negative." When neither of these is appropriate, either repeat the instruction along with your intentions or use a word that does that for you. Study the use of these terms and use the appropriate answer the first time. Additionally, always use your call sign with each transmission so that we know who is speaking.

As difficult as it is to deal with useless information, it is much more difficult and potentially dangerous trying to deal with erroneous replies or replies by the wrong pilot. There are several built-in traps in this area about which we all have to be very aware. Let's take a look at some of the most common.

CALL SIGNS

The first of these is the problem that all facilities have with similar sounding call signs. When we built our fictional aviation empire in the introduction of this book, we set up circumstances which are similar to real life corporations. We created the potential for misunderstanding by giving our aircraft similar sounding call signs. Companies certainly do not try to confuse the controllers intentionally. It is simply a case where a good marketing strategy, and the strategist who created it, does not take into consideration the potential traps involved in that idea.

Tandem Aviation, which is a very good company, likes to purchase corporate and training aircraft with call signs that end with the alphabet characters of their corporate name. They also like to arrange their trip numbers in a logical sequence based on the type aircraft and the direction of the routes that they fly. Since their corporate aircraft are usually based at the same airport and their airline trips are scheduled at the peak travel times, it is not unusual for several of them to be on the same frequency at the same time.

Let's say that Tandem Airlines marketing technique is to create a call sign with the first digit representing the type aircraft. For example, all flights in a DC-9 will have 900 numbers. They then break down the trip number further to have the second digit generally represent the direction of flight. For example, east/west operations will be odd numbers and north/south flights will have even numbers. So an eastbound DC-9 could have a 950, 970, 990, etc. series number. The third digit will distinguish one aircraft trip from another and is often representative of the number of trips that have been taken on that route that day.

Since most people like to travel to a destination airport in time to arrive at the beginning or end of a business day, these times tend to be very busy. (We call them "rushes" or "pushes.") On any given arrival rush it would not be unusual for a controller to be working TA998, TA988, TA989, TA789, and TA978 at the same time. Add to this the fact that some airlines like to give their international flights just two digits, and we may also be talking to TA98. Just so that you will feel sorry for the poor controller, let's add N98989 on the frequency along with everyone else.

This problem of similar sounding call signs is an area where even the most experienced pilots can make an error, and it is fairly obvious that a very busy facility, with a lot of aircraft with similar-sounding call signs on a frequency, creates a potential for a pilot to take an instruction intended for another pilot. However, this same potential exists when there are only two aircraft on the frequency. If these two aircraft have similar-sounding call signs and there are long periods of silence on the frequency, the pilots begin to expect to hear a transmission directed at them and may miss the subtle differences between the call signs.

This situation is further complicated by the fact that most facilities use standard techniques in vectoring aircraft to their destination. What this means is that each pilot will receive basically the same instruction at the same point in his/her flight. Descent instructions, turns, and clearances will occur in a logical sequence that will be the same for every flight. Pilots who fly these routes become accustomed to receiving these instructions in that sequence. As you can see, the stage is now set for a pilot, expecting a clearance, to hear one.

If you, as a pilot, are not paying close attention, or if two aircraft with similar-sounding call signs are on two different frequencies being worked by the same controller, it is entirely possible that the wrong pilot will act on a control instruction intended for someone else. Remember, you only hear the responses from the pilots on *your* frequency, while the controller hears both frequencies.

This can be a double-edged sword because the controller is expecting to hear a reply that matches his/her transmission. He or she may miss the slight difference between what is expected and what is actually said. Additionally, if the pilot response was blocked, by the time the controller determines who said what, the wrong pilot may have already started the descent or the maneuver.

As I indicated earlier, the above scenario is usually, but not always, found at the larger airports with their higher volume of air carrier traffic. Smaller facilities are by no means immune to this type of communication problem. Let's suggest that Tandem Aviation has a big flight training facility near a large college in Medianville, U.S.A. Just to make it difficult we will construct our scenario so that this control tower does not have radar to extend the eyes of the controller. This flight training school, like the one where you may have learned or are now learning to fly, is a professionally operated flight school. The staff at this organization tries to schedule the students for a full hour of flight training with adequate time added for a good preflight and postflight briefing.

So every 90 minutes, one dozen red-white-and-blue Cessna 152s taxi for takeoff. Their pilots, all using call signs that end with Tango Alpha, call the control tower for taxi instructions. Can't you just imagine the controller trying to tell an inbound aircraft to follow the red-white-and-blue Cessna on downwind, while chewing on the last antacid tablet in the bottle. Just for the fun of it, we can schedule the annual widget manufacturers trade show at the local convention hall and add a few corporate Tango Alpha Learjets full of petrified millionaire MBAs to mix in with the Cessnas.

In reality, the number of times that a pilot will take instructions intended for other pilots is fairly small. But when you consider the tens of thousands of control instructions and replies that fill the airwaves each day, I think you can see that the potential for these incidents is high. I would like to be able to tell you that, in every case, you should not act on the control instruction unless you check with the controller to make sure that it was intended for you. But occasionally, this lack of action can be just as bad as acting on the wrong instruction. Not acting is certainly better in most cases, because a smart controller will not allow a situation to develop where immediate compliance is the only thing separating their aircraft.

PAY ATTENTION

The only real solution to these potential traps is for every pilot and controller to pay strict attention to what is being said on the frequency at all times and save the social chatter for the airport lounge or break room. This philosophy is particularly relevant in the training aircraft. Instructors like to teach when the situation is fresh in the mind of the student, and some will even turn down the radio so they can be heard. (This is when the controllers dream about machine guns mounted on the top of the tower cab.) I have been in these situations and the desire to teach instead of listen is very strong. But there are several things that both the student and the instructor should keep in mind.

First, the student in this situation will either concentrate on the instructor and forget the flying, or tune out everything (including the instructor) to concentrate on the flying. This means that he or she is not listening to the radio. Second, the instructor's ability to listen to the radio is reduced while talking or eliminated if the radio has been turned down. These two facts probably contribute to the third, and perhaps most important, reason for listening to the radio in a training environment: FAA statistics reveal that approximately 50 percent of all midair collisions occur near an airport and involve aircraft that are conducting some type of training. These aircraft generally contain a flight instructor and often another pilot who has experience above the private pilot level. This suggests that one or both of these pilots was not paying attention to what was going on around them. I do not intend to get into analyzing the cause of accidents, that is for organizations such as the National Transportation Safety Board (NTSB). If you would care to read some of their reports, I think that you will be surprised at how often a communication error is a contributing factor.

GREAT EXPECTATIONS

A little while back I mentioned the fact that we sometimes hear what we expect to hear over the radio. This is the type of error that doesn't require a busy situation with complicated instructions and similar sounding call signs to jump up and bite us. These errors occur among pilots and controllers regardless of the level of traffic. Those involved are sometimes experienced, conscientious, professional pilots or controllers who are familiar with the routes, procedures, and the area where they work. So let's analyze some of the things that could be behind this type of error.

The use of standard instrument departures (SIDs) and standard terminal arrival routes (STARs) gradually begins to lead pilots and controllers into the habit of expecting each other to do what is routine. Sometimes just the everyday use of the same type of control techniques in the same types of situations has this same effect. This works fine until one of the participants has to do something unexpected. If a pilot is used to operating out of an airport where the SID requires an aircraft to level at 5000 feet, and he then goes to an airport where the SID stops the climb at 4000 feet, the pilot must make a conscious effort to change a habit pattern.

On the other side of the radar scope, controllers occasionally find that the traffic situation requires a change from a standard procedure to one that meets the individual requirements for that moment. Unless the controller emphasizes the fact that this is a change, there is always the chance that the pilot, who is expecting a routine transmission, will miss that change.

Another problem is that some pilots know the system well enough to expect a particular instruction at a certain point in their flight. When they hear that instruction they automatically assume that it was for them. This leads to some real surprises when two aircraft, or the wrong aircraft, executes a maneuver at a time when only the correct aircraft will fit. A considerable portion of a controller's time is spent checking the execution of his/her instructions and reacting if those instructions do not work.

Individual aircraft peculiarities sometimes catch controllers off guard because they too are expecting routine performance. If an aircraft climbs slower or faster than normal, has to level off to retract flaps, or is operating at reduced performance because it is very heavy, the controller may not recognize this fact until it becomes a problem. These kinds of problems are among the most insidious because neither the pilot nor the controller is expecting anything unusual. The controller is conducting a routine traffic scan and does not give that flight special attention, and the pilot is confident that the maneuver is in accordance with the controller's wishes. This is one example of how it is possible for situations to creep up on us without anyone being aware of what is happening, but this is not usually the case. Most times, someone somewhere senses that something is wrong.

PROFESSIONAL SKEPTICISM

In 1986, Atlanta Tower manager Harry McIntyre published a Letter to Airmen (LTA) titled "Professional Skepticism" which deals with our mutually shared responsibility to identify and correct errors. Its basic premise is that we *all* share a responsibility to ensure that the air traffic system remains safe:

> Safe, orderly, expeditious—the watchwords of the U.S. Air Traffic Control System. These words, from the beginning of ATC, have carried the missionary message to the controller and pilot on priorities and purpose.
>
> Our ATC System has produced many successes as it has evolved from its rudimentary beginnings to today's sophisticated technology. The most important ingredient to this success however, has not been technology but the cooperative relationship between pilots and controllers communicating with one another—exchanging data, making judgements, executing decisions, *and yes, correcting errors committed by each other.*
>
> The person in the best position to identify and correct human errors is obviously the person who made the error. But as we often say—"to error is human." I suspect we say this because we tend to overlook our own performance or see it better than it actually is.

> In the ATC System, there's another person, either the pilot or controller as the case may be, who can identify and correct the error committed by the other. This does not mean pilots and controllers need to assume an adversarial relationship or even be each others' critic—rather, it means we must "question" when we are in doubt about any aspect of the operation at hand. As a term of reference for this need, let's call it "Professional Skepticism."
>
> "Professional Skepticism" is objectively reviewing the operation at hand and questioning or seeking clarification when communications or events don't appear right. As professionals, we must be sure of our data and therefore question and seek clarity whenever we are in doubt.
>
> A review of several incidents reveals that too often we avoid the direct method of seeking clarity and opt for the indirect, vague, style. We probably tend to do this in deference to the other professional. For safety sake, we *must avoid* this *"gamesmanship"* and *be a "Professional Skeptic."*
>
> In summary, the practice of "Professional Skepticism" can have a significant impact in improving our already enviable safety record in the ATC System. As *professional skeptics*, pilots and controllers must always:
>
> - Seek clarity when instructions or information are not clear.
>
> - Question an operation that does not seem right—but say *why* it does not seem right (be specific).
>
> - Follow prescribed procedures and phraseology in our voice communication technique.

This letter focuses on the key to preventing, recognizing, and correcting these types of problems: communication among all parties involved. This involvement is not limited to the pilot and controller who are talking to each other. Sometimes the person who knows that something is wrong is not one of the people who is directly involved in the exchange. It could be any other person on the frequency who heard the exchange and is aware that a problem is developing. This knowledge makes you an involved party and you are morally responsible for doing something about that knowledge.

If you think back over your flying career I'm sure that you can think of at least one incident where you heard a radio exchange that you knew was misunderstood or was answered incorrectly. Did you say anything about it to the participants? Probably not, because you felt that it was going to be corrected by the controller or pilot involved, or you refused to allow yourself to believe what you thought you heard, or you simply did not want to become involved. We are not suggesting that everyone second-guess each transmission, but if you honestly believe that the exchange was in error or it just "felt wrong" to your instincts, say something. If you don't want to look foolish in the event you are wrong, make a blind broadcast that describes what you think was wrong. An example would be, "XYZ Approach, I think Tandem 998 read back the wrong altitude" or "Approach, I think the wrong aircraft responded to your last transmission." This action will trigger an immediate series of exchanges to verify the correct intent on everyone's part and the worst that will happen is that you will be proven wrong. If you are right, you will have prevented a potentially dangerous situation from developing and you may even get a blind broadcast "thank you" from one of the participants. It will also impress the other members of your crew and you will feel great the rest of the day.

BODY LANGUAGE

The types of communication that we have been describing so far are some of the ways that information can be verbally transferred from one person to another. Other forms of communication are more subtle and may be nonstandard in the sense that they are not published as procedures to be taught routinely. But these techniques are just as effective in their ability to transfer information and are often the only methods that are available to an individual in a particular situation. Let me give you an example of how good pilots use nonverbal methods to reinforce the fact that they have received, understood, and are complying with control instructions.

A large number of professional pilots have developed a technique for use when taxiing at night, involving the use of landing or taxi lights, that controllers have found very helpful. When a controller issues an instruction to a pilot to hold short of an intersection or runway along the aircraft's taxi route, that controller visually checks to make sure that the aircraft stops at the assigned point. This is much more difficult at night because most people have a reduced sense of depth perception in the dark. This loss of ability is further aggravated by the distances that we have to scan and the large number of lights that are operating around an airport. Because of this fact, most controllers find that, when it's dark, they have to be much more aware of the *exact* location of taxiing aircraft. The result is that they more frequently ask for verification from the pilot that the aircraft is going to hold short as instructed. Often this extra check is done because it appears that the aircraft is coming too close to the intersection. We realize that this is irritating to pilots but we just cannot take a chance that the aircraft might not see the hold lines.

It is at this point that the use of nonverbal communication techniques can be used to pass information to the controller. When an aircraft reaches the hold lines or intersection, especially those at a runway crossing point, many pilots will turn off their landing or taxi lights. This, combined with the pilot's verbal acknowledgment, informs the controller that the aircraft has stopped. After the controller clears the aircraft to continue taxiing or to cross the active runway, the act of turning the lights back on reinforces the fact that the pilot received the transmission. Aside from the fact that this technique is a form of courtesy to other pilots by keeping the lights out of their eyes in crossing operations, it also serves as a subtle form of communication to the controllers that information is being transferred.

FLIGHT PLAN FLUBS

Let's move away from the direct pilot/controller forms of communication for the moment and go back to the fact that there is frequently a misunderstanding among pilots about the types and amount of information that controllers have regarding pilots' flight plans. This is also a type of information transfer and a breakdown in this process is an equally frustrating communication breakdown.

When a pilot files a flight plan with FSS, they give the specialist information regarding the route, altitude, type of aircraft and equipment, pilot's name, color of aircraft, and several other items of information that are nice to know. The pilots assume that all of this information follows them throughout the system and that they are expected to provide this information when transmitting a request for an IFR clearance directly to the controller. Unfortunately, both of these assumptions are incorrect.

In the first case, some of the information resides with the FSS as a form of historical data and is not included with the information transmitted to the controller. We do not know the color of your aircraft or, unfortunately, the name of the pilot. Additionally, given the thousands of flight plans that are filed every day, some erroneous information might find its way into the system and provide some real surprises for all concerned.

When a pilot inadvertently gives an incorrect fix or misspells an identifier, the computer will, in most cases, refuse to accept the flight plan. The FSS specialist is then forced to guess the correct route and enter a code on the flight plan that requires the controllers to issue a full route clearance as a form of verification. If the pilot files for a route that includes direct navigation, the computer will generally accept that as a valid route whether the aircraft is equipped to fly that route or not.

We frequently see IFR flight plans that show routes from point A direct to point B flown by aircraft that contain no area or inertial navigation equipment. We normally vector these aircraft clear of arrival or departure traffic routes then tell the pilot to "proceed via present position direct destination airport." The typical pilot response is a request for a heading until he or she is able to do that. If the destination is not depicted on our video map, the best we can do is aim the aircraft in the general direction, pass that information to the next facility on route and wonder what kind of understanding this pilot has of the system.

The other case, where the pilot requests an IFR clearance directly from the controller, offers some rather unique situations and is often frustrating to the pilot.

When a pilot tries to airfile directly with the enroute or approach controller, there are usually two factors which determine whether the controller will issue the clearance or require that the pilot file the flight plan with the FSS. First, if the destination airport is within the air traffic control facility's area of responsibility, the controller will almost always comply with the request for a clearance. Second, if the airport is not in their area of responsibility, the controller has to have the time to take away from the duties of separating aircraft to perform the mechanics of filing that flight plan. Let's examine both of these situations more closely.

Who Owns What?

Each air traffic control facility has a designated amount of airspace for which it is responsible. That area is dependent on the size of the facility, its radar and radio coverage capability, and its proximity to other larger air traffic control facilities. A specific ATC organization may own only a few square miles of airspace at low altitude or tens of thousands of square miles up to the highest flight levels. In the Atlanta area, for example, Dobbins AFB owns a small ground controlled approach (GCA) pattern of only a few square miles around the airport at 3000 and 4000 feet MSL. Atlanta approach control owns, with the exception of airport traffic areas and the delegated GCA pattern, all of the airspace within 40 NM of the Hartsfield airport (ATL) from the surface to 14,000 MSL. Atlanta ARTCC, on the other hand, owns all of the airspace not delegated to approach control facilities over a large portion of the southeastern United States.

If a pilot, requesting a cruise altitude of 14,000 feet or lower, departs the Carrollton airport (35 miles west of ATL) wishing to fly to the Monroe airport (37 miles east of ATL), a distance of over seventy miles, an Atlanta approach controller can issue an IFR clearance to this pilot without further consultation with any other ATC facility. This is because the entire route lies within Atlanta's airspace. All the controller has to do is enter the aircraft call sign into his ARTS keyboard and handwrite a flight progress strip for the operation. The *only* information the controller really needs to accomplish this is the aircraft's call sign. (FARs 91.83 and 91.115 describe pilot requirements for filing a flight plan. This information, together with the techniques that are actually used in flight plan filing, will be covered in more detail in Chapter 5.) The controllers will, of course, ask for additional information. We usually ask about the type aircraft and the equipment capabilities so that we can plan for the separation requirements, know what types of approaches the aircraft is capable of conducting, and have some idea of what performance capabilities to expect from that aircraft.

Let's change the situation around a bit and I will show you how a slight modification in the pilot's request can make a major difference in how it is handled. If the same pilot wants to fly from Cedertown (37 miles NW of ATL) to Rome

(48 miles NW of ATL), a distance of less than 20 NM, the situation is entirely different. The Atlanta approach controller cannot issue an IFR clearance without obtaining a significant amount of information from the pilot and conducting some type of coordination with Atlanta ARTCC, in whose airspace the Rome airport lies.

These are specific instances relative only to the Atlanta airspace, but every air traffic control facility has similar airport placements that meet the same sets of circumstances. Pilots frequently think that just because they are only going a short distance they can kick the tires, light the fires, and call approach control for a short-range clearance. Somewhere along the way they have also heard the phrase "Tower Enroute" which refers to a flight between two approach control facilities which have a common boundary. Pilots think that, since their destination is within the next approach control facility's airspace, all they have to do is ask the controller for an IFR clearance and they can expect to be handled just like the pilot going from Carrollton to Monroe.

These pilots are making two mistakes that are, unfortunately, typical of the way many pilots operate in the air traffic system. First, they do not know who owns the airspace in which they fly. Second, they do not know the process required for the controller to enter a flight plan which will allow them to fly within that system.

The first of these errors has a very simple solution. Virtually any pilot group can initiate a request to the local air traffic facility for a speaker who will explain the airspace configurations in use in their area. Most facilities have prepared presentations and designated personnel who handle this type of request on a routine basis and we welcome any chance to further the understanding between pilots and controllers. We can also arrange for a group of controllers to attend a pilot/controller forum for a question-and-answer session with no preset agenda. Additionally, group tours of the ATC facility can be arranged so that pilots can actually see the air traffic operation. The National Airspace System is similar to a multilayered jigsaw puzzle. The total picture cannot be seen until you know where all of the pieces fit and then put it together. Like that puzzle, when you only know what one piece looks like, you are reduced to guessing about its location in the total picture.

Airfiling IFR

The second error is a little more difficult to eliminate. The many steps necessary to enter a flight plan into the system involve considerable frustration for the controller. Only by sitting down and watching a controller go through this process, can a pilot understand why controllers are so quick to send him/her to the FSS frequency to file a flight plan.

Another thing that the pilot doesn't understand is that the process of filing a flight plan with the FSS takes longer than the time it takes to express their desires into the microphone. Because of the physical work involved in the actual typing of the flight plan into the computer, the time involved in looking up identifiers, and the time required for the computer processing and issuing of the flight plan according to preset priorities, it is not unusual for 10 to 15 minutes to pass before the actual flight plan is available. During this time, the pilot is in the air, usually not very patient, and is often muttering to himself (and sometimes the controller) about the inefficiency of the system.

Since a controller cannot authorize an IFR aircraft's penetration of another controller's airspace without detailed coordination or a valid flight plan and handoff, he will not issue a clearance that turns a VFR aircraft into an IFR aircraft until such a flight plan exists. Often the aircraft passes out of that controller's airspace jurisdiction before this can be accomplished, and the controller must have the pilot contact the next controller enroute while the pilot waits for his flight plan to process. Or, if the pilot is encountering IFR weather and cannot continue, the controller might have to issue a short range clearance to a point within his airspace where the pilot will hold while awaiting the flight plan.

Most people reading this would think "I would never do something like that," but you would be surprised at the circumstances that arise which cause normally conscientious pilots to do things that make a controller see red.

Now, supposing that the controller does have time to file a flight plan for the pilot, let's talk about what information he or she needs to accomplish this task. When a pilot asks to file an airborne IFR flight plan to a destination outside the controller's airspace, the controller only needs to know the following: The aircraft call sign, the type aircraft and the equipment suffix, the aircraft speed, requested altitude, the origin point of the flight plan route (since the pilot is talking to the controller we can pick a point in our airspace from which to begin the route), and the complete, accurate route of flight.

Unfortunately, because of FAR requirements for obtaining a clearance under IFR, many pilots believe that they have to read to the controller, on the frequency, all of the information that they would normally give the FSS specialist. (Unless a controller authorizes the elimination of some elements, this belief is actually correct. We will cover this in more detail in Chapter 5.) This takes time, contains a large amount of information that the controller does not need, and pilots frequently leave out a needed element of the information. Usually pilots forget that someone is trying to write this down and they talk so fast that the controller misses part of the route and has to play twenty questions with the pilot to obtain the correct data. In most cases they also read the route elements by their full title instead of the appropriate three- or five-letter designator (Collier VOR instead of its designator, IRQ) and the controller has to look up those identifiers unknown to him or ask the pilot (who frequently doesn't know them either) what they are.

Once all of the information has been gathered, the controller has to type the information into the computer using a keyboard. This particular keyboard is a separate piece of interface equipment that is designed to work with the mainframe computer in use by the ARTCC in whose airspace the operation is being conducted. Most FAA facilities currently use a system called *flight data encoding printer* (FDEP), or the newer generation, *flight data input/output* (FDIO) that is connected to the ARTCC's host computer through some type of *data control unit* (DCU). The DCU establishes priorities based on workload and type of message and will not process a flight plan until all higher-priority messages have been transmitted.

The FDEP serves the data requirements of all positions within the facility and, as a result, is generally positioned in some central location in the radar room or tower. Because of this, it may not be within reach of a particular radar scope or controller. What this means is that any controller who wishes to enter a flight plan into the FDEP has to leave the radar display to do so or has to find another person and relay the information to them. If a controller is busy with a complex traffic operation (which does not necessarily mean a large amount of traffic), he or she is usually unable to spend the amount of frequency time necessary to gather the information required for filing the flight plan. Even if they could find the time, it is generally not possible to leave the radar scope to file the flight plan.

When pilots visit our facility, they are often amazed at the amount of time that a controller spends performing duties other than talking on the radio. It is not unusual for the controllers to make one or two calls on telephone lines coordinating with other controllers, enter information into the ARTS keyboard, and hand flight progress strips to another controller between transmissions to pilots. Pilots hear this dead airtime and think that the controller is doing nothing and is being arbitrary by refusing to take their flight plan.

Controllers occasionally have another person working with them at the radar position who functions as a handoff or coordination specialist. This individual could enter the flight plans into the FDEP since they usually do not have direct responsibility for a radar display. However, a handoff position often has coordination responsibility for more than one controller and is usually busier than the proverbial beaver. So, even though they may have the ability to do that job, they have even less time to perform this task than does the radar controller.

The location of the aircraft when the pilot asks us to enter a flight plan is often a factor in our decision to comply with or decline the request. If the aircraft is only a few miles from our airspace boundary, it is unlikely that the filing process could be completed before the aircraft passes into another controller's area of responsibility. Beyond the fact that this would involve a significant amount of coordination between two busy controllers, airspace boundaries are usually close to the limits of a controller's radar coverage area. To complete this operation would involve issuing an IFR clearance to a pilot in a nonradar environment in another controller's airspace. This is unsafe, unwise and, unless coordination defining the responsibilities of all involved parties is completed, it is also a violation of ATC procedures.

When conducted improperly, this type of action is called a *systems deviation* (a violation of another controller's airspace without proper coordination) and the immediate result is that the controller is decertified and is unable to work traffic until the incident is investigated and steps are taken to ensure that it does not happen again. If this deviation also results in less-than-standard separation existing between the first aircraft and another aircraft in the other controller's airspace, it is upgraded to a *systems error* (a loss of minimum separation) which is considerably more grave.

The FAA is dedicated to conducting a ''*safe*, orderly, and expeditious'' operation with the emphasis on safe. Any controller who establishes a history of deviations and errors may be invited to exercise other career options. Given these circumstances, plus the nagging doubt about what type of preflight briefing this pilot may have had, a controller could be betting his or her career on this type of operation. I'm sure you can understand why we would be reluctant to jump every time a pilot wants to airfile.

Not So Fast

There is one other erroneous type of operation connected with flight plan filing which results in a similar set of circumstances and is often a source of irritation to all concerned. Let me give you some background information about the mechanics of the flight data processing system, and then take a look at what happens when Fast Freddie Flier files a flight plan.

When a pilot files a flight plan, that information reaches the controller through a series of computer exchanges based on the proposed departure time of the flight. The FDEP system prints out the flight plan information approximately 30 minutes prior to that time. At the same time, the information is routed to the appropriate radar display's ARTS storage list (called a *tab list*) which is capable of displaying 26 tracks. Depending on the workload of the computer and the number of flight plans scheduled for that same time period, the track may not be displayed in the ARTS tab list during this 30-minute time period, but it is accessible through a series of keyboard entries.

The flight plan may also be accessible from the central computer prior to the 30-minute printing time but the computer will not have assigned a transponder code to the flight before this time. Again, a series of keyboard entries and responses is required to move the information into the local computer. Since it takes about 15 minutes for the action of filing a flight plan to be accomplished, the controller has no information concerning the flight until the process has been completed and the host computers have transferred the information to the local computer system.

If Fast Freddie files his flight plan and then immediately jumps into the air, or if he departs more than 30 minutes prior to his proposed departure time, the controller will not have any information concerning the flight. In the first case, there is absolutely nothing the controller can do until the flight plan is processed and entered into the system. An attempt to enter another flight plan into the system could, depending on the timing of the entries, result in duplicate flight plans and the wrong flight plan being activated. We've already talked about the problems that can cause.

In the latter case, the very least that will be involved is an increase in the workload of the controllers who have to amend the proposed departure time and request the flight plan from the computer. In either case, the aircraft is in the air, unable to obtain an IFR clearance, and occupying a considerable amount of at least one controller's time and resources because of poor preflight planning.

What I have tried to demonstrate in these last few pages is that a breakdown in communication does not necessarily have to involve a misunderstanding between two individuals talking directly to one another. Rather, it is often the misuse, incorrect application, or failure to use the equipment that is increasingly being relied upon to transfer information between pilot and controller. Many pilots don't realize how much information a controller is required to give to them or get from them during a routine operation and, consequently, how much time this would involve if you multiply it by the number of operations we are conducting.

USING AND ABUSING THE SYSTEM

Unfortunately, the equipment and procedures that we currently use are not being used correctly and, unless we can teach pilots to properly use them, the future application of new complex equipment and procedures does not look promising.

Let's look at how pilots currently misuse or fail to properly use services—ATIS and Notices to Airmen (NOTAMs) for example—and show how this throws a monkey wrench into the ATC system.

ATIS

ATIS is simply a tape-recorded statement that is broadcast over a frequency set aside only for this purpose. The recording is usually 30 to 60 seconds long and contains information regarding the current weather, the runways and approaches in use at the airport, specific unusual conditions such as wind shear, braking action, bird activity, or runway and taxiway closures. It also advises about equipment outages and gives general instructions relevant to that airport's operation. Each new recording is identified by an alphabet code which changes sequentially from A to Z when an ATIS is updated. For example, a tape might begin by a controller saying "This is Medianville Airport Information Alpha."

Some larger facilities use two ATIS frequencies and assign one for arrival aircraft (ATIS codes A through M) and the other for departure aircraft (ATIS codes N through Z). A lot of pilots complain that these broadcasts are too long and contain too much information and that they have to listen to them twice to be sure that they hear all of the information. What they fail to realize is that if ATIS did not exist, a controller would have to broadcast that information to every pilot who called on the frequency. At a place like Atlanta, that would result in an additional 2500–3000 30-to-60-second broadcasts each day. That works out to between 20 and 40 hours of controller talk just to accomplish what each pilot spends two to three minutes doing.

Time management in the cockpit is very important and a lot of professional pilots find themselves in a position where they are switched to the approach control frequency before they have had time to listen to the ATIS and give it the attention that it really deserves. Similarly, the VFR pilot, who is a little behind the aircraft or the flight plan, will suddenly find himself near his destination airport sooner than anticipated and will not have listened to the ATIS when it is time to call the control tower or approach control facility. Often, pilots will give only a cursory listen to ATIS—just enough to learn the runway in use and tell ATC the current ATIS code. Some pilots only listen long enough to hear the code, because they know that the controller will tell them the approach for which they are being vectored and/or the runway on which they will land.

I don't know whether pilots do this because they think they're beating the system. Perhaps they think they're too important to waste time doing things they believe the controller should be doing. Or maybe they just don't realize how critical this information can be. Unfortunately, part of the reason they continue using this technique is that they can usually get away with it, and the habit becomes part of their normal operating practices. Some pilots will report on the frequency and not give any indication that they have the current ATIS code. Controllers frequently hear such phrases as "we've got the numbers" or "N50TA is with you with the current ATIS." This *always* involves an additional workload on the controller, who must determine that the pilot has the correct, current ATIS. If not, the controller must read specific parts of the required information to the pilot so that there will be no doubt that the information was received. I sincerely doubt that these same pilots would accept an equally incomplete briefing from their scheduling department, the FSS specialist, or their flight instructor, but they don't apply the same requirements to themselves. When a pilot offers this type of initial exchange to the controller, he has made no friends at the tower and we will begin wondering whether he will do other equally sloppy things.

Even when the pilot, who has not taken the time to listen to the whole ATIS, reports the correct ATIS code, (he heard it from the pilot who called in just ahead of him) it usually does not take long for it to become apparent to the controller that this pilot is blowing smoke from more places than just the engines.

Let me relate just a couple of generic "war stories" that make the point:

TA998 reports on the frequency and advises that he has ATIS Information Alpha. Alpha is current and it advises the pilot to expect vectors for a visual approach to Runway 23 and informs him that the Runway 23 ILS is out of service. The controller tells the pilot to expect vectors for a visual approach to Runway 23 and begins those vectors.

Several minutes later, after the controller has cleared the aircraft for a visual approach, the pilot says, "Approach, we're not receiving the localizer."

The controller, while observing the target go across the final approach course toward another aircraft, thinks, "There is a good reason for that, Captain," but only responds, "Yes sir, the ILS is out of service, turn right heading 250 and proceed direct to the airport."

Some pilots will even carry this situation to the point of chewing on the controller for his failure to provide this important piece of information.

The controller thinks, "what a turkey," and responds, "Sir, the outage was reported on ATIS Information Alpha which you indicated that you had, contact the tower."

I know that the pilot must feel kind of small when this happens, so do the job right the first time. Don't start chewing on something until you're sure that it's not crow that you're going to be eating.

Like any other tool, the ATIS is only useful when it is used properly, and contrary to popular belief, the ATIS really has two main functions. First, it is designed to reduce the workload on the controller, and for that we are eternally grateful. Second, and perhaps most importantly, it allows the ATC facility to provide the pilot with more complete information than we would ever have time to broadcast over the frequency. If just one warning of potential wind shear, broadcast on ATIS, causes just one pilot to carry the little bit of extra airspeed to save his aircraft, then the extra workload of carrying a handful of lazy, inattentive, and "me first" pilots is worth the effort.

NOTAMs

The Notice to Airmen (NOTAM) procedure is much like ATIS in that it is designed to provide the pilots with advance information about the area or airport(s) to which they are flying. Perhaps the greatest comedy of errors, and best example of poor pilot planning I have ever seen, involved the misuse of both of these pilot services.

The pilot of N123TA called the XYZ approach controller and informed her that he had the current ATIS information for one of the satellite airports under her jurisdiction. He went on to advise that he wanted to land and refuel and then conduct a series of practice ILS and VOR approaches to the airport. I believe that the controller was beginning to think that this was some kind of a joke. The ATIS at his destination airport was showing that the airport had been closed for over an hour and would remain closed for another two hours due to an airport open house and airshow with high-speed aerobatic aircraft. Additionally, the VOR that had served this airport's VOR approaches had been decommissioned for more than six months and would not be moved to its new location for several more months.

When the controller related this information to the pilot, he was incensed that he could not land at the airport and even suggested that the airport be reopened so that he could land. When the controller advised the pilot that this would not be possible, the pilot's next response, a request for vectors to the nearest airport, really got her attention. She responded by giving the pilot all of the information regarding nearby airports and offered vectors to any one of them. She also asked the pilot for information regarding the fuel status of his aircraft because she suspected he may be low on fuel. At this point, the pilot's responses became evasive, sometimes sarcastic, and generally not helpful to the controller who was trying to help him. He indicated that he didn't have time to fool around waiting for people to quit playing games and wanted to get to another airport without delay. When the pilot was vectored to another airport, an uncontrolled airport and the closest one available to his original location, he abruptly terminated our services and left the frequency to land.

From this scenario we know that the pilot did not listen to the ATIS and did not check the NOTAMs or he would have known that the airport would be closed and the VOR had been decommissioned. From the pilot's actions and statements, I suspect that he did not plan to have enough fuel on board his aircraft to travel very far beyond his original destination airport. We can also surmise that the pilot did not obtain a briefing from the FSS specialist because that individual would have informed the pilot of the temporary NOTAM regarding the airport closure.

NOTAMs are published in several different formats depending upon the importance of the item being reported and the length of time that the information is relevant. The NOTAM L is a class of NOTAMs which is restricted to local dissemination and usually is relevant only to the facility in question. An example of this would be the closure of a taxiway on an airport. When a change in equipment or available services has a more wide-ranging impact, it is published as a Class D NOTAMs. The information associated with this NOTAMs would be included in the weather sequences issued for that airport and/or the general briefing issued by the FSS specialist for pilots operating in the affected area. As an example, the pilot operating into the airport that was closed by the open house and airshow would have received a briefing on this Class D NOTAMs. Generally, these Class L or D NOTAMs are of short duration or limited impact to the general flying public (limited in terms of affecting a relatively small area when compared to the whole country).

When the change will be of a longer duration, it is included in the next scheduled publication of the Class 2 NOTAMs handbook. This handbook contains information about changes and outages that will be in place for more than seven days. If these changes are of a permanent or semi-permanent nature, such as a change in the airport lighting systems, or long-term or permanent runway closures, they will be published in the *Airport/Facility Directory*, which outlines the services, equipment, and navigational capabilities at each airport.

COMMUNICATION—CORNERSTONE OF ATC

The use of NOTAMs, ATIS, and other informational services available to pilots, such as the *enroute flight advisory service* (EFAS or "Flight Watch") and *transcribed weather broadcasts* (TWEBs), significantly reduces controller workload by eliminating the need to broadcast this information over a busy frequency. A pilot's failure to use these informational services, failure to properly state what he means, or failure to understand what is said when talking to the controller, usually results in a communication breakdown. While most of these are resolved by a series of exchanges between the two parties, the fact remains that this always results in an extra workload on the controller, and it has the potential to be embarrassing, costly, and even dangerous to the pilot.

If you are a flight instructor, consider how much credibility is lost with your student when you fly somewhere to conduct approaches only to find out that the approach is NOTAMed off the air. You spend a lot of the student's gas, money, and time to get there and accomplish nothing. Or you may place yourself in a situation, like the airshow example earlier, where unexpected high-speed aerobatic activity could place the safety of your aircraft in jeopardy. When I hear flight instructors talk about students who seem to jump around from one instructor to another, I wonder how many of those students have actually fled for their lives.

Communication is one of the cornerstones of the air traffic system, and the pilot errors associated with the failure of this fundamental are cited as a cause in over half of the accidents investigated by the NTSB. In a large number of the pilot deviations handled by the FAA's flight standards division communication errors are cited as a contributing factor. The "human errors" that cause most of these events are often the result of bad habits that have been formed over a period of time. Every pilot has the responsibility to recognize and eliminate these habits and conduct his or her operation with a high degree of professionalism.

Stay out of the airplane until you have become familiar with every possible aspect of your proposed flight from every source available. When you climb inside, treat the radio as an extension of your mind. It provides you with the only method by which you can "see" what is going on around you and get the "feel" of where you fit into the total picture. Pay attention to what this extrasensory device is telling you, and make sure that the information corresponds with what all of your other senses are telling you. Be careful not to allow your expectations to mask reality. Don't let distractions cause you to miss important pieces of information. Most importantly, do not conduct an operation unless there is a reasonable certainty that you are doing what is right. If ATC instructions seem wrong for the situation, ask for a clarification. Always leave yourself an out, because other people make mistakes too. Remember, when in doubt, check it out.

3

Phraseology and Word Concepts

WHEN YOU SPEND THE BETTER PART OF YOUR ADULT LIFE LISTENING TO PIlots and other controllers talk over a radio and then compare this to what the general public perceives to be aviation ''lingo'' (as derived from the various movies, newspaper stories, and television), you begin to get a feel for how special our language is and can understand why most of us jealously guard its uniqueness. Virtually every word we use has some important connotation, and the way we use or combine those words often carries a message that is far greater than just the sum of the words. It is extremely important that all members of the aviation community understand that some aviation terminology, words, or concepts have specific meanings, applications, and/or limitations. When pilots don't understand the meanings of these words and phrases, when they have incorrect expectations of the actions associated with a particular word or term, when they substitute terms that have no meaning in aviation, or when they misuse a phrase or a word within the context of their transmissions, the results can vary from frustrating to disastrous. I can remember one situation from my own experience that illustrates this frustration and also demonstrates the pride that some pilots seem to have with regard to our language.

In the late 1970s, when the CB (citizens' band) radio craze was sweeping the nation, I was stationed as a controller in Michigan. One afternoon a pilot called in on the approach control frequency and said that his ''20'' was somewhere south of the airport and he needed to ''10'' something or other at the ''airpatch.'' This nonstandard phraseology was irritating, but the only response that I could make was to tell the pilot that I did not understand his transmission. After the pilot transmitted another round of CB jargon, I responded with a transmission designed to indicate what I understood of his request. To this the pilot responded, "10-4 good buddy."

Before I could say anything else, another pilot jumped in. "Listen dummy, that controller is politely trying to tell you to use the correct phraseology. If you think that garbage is cute, park the aircraft, drive a truck, and quit wasting everybody's time."

The pilot who chastised this individual probably did not have a very high opinion of this language but, regardless of personal opinions, part of his suggestion was still appropriate because it did point out the need for the use of correct phraseology. The use of CB language would have been perfectly acceptable had that individual been using it in the environment for which it was designed and had he been dealing with someone who understood what he was saying. Unfortunately, I was not conversant with CB language and, even if I did understand what he was trying to say, I would not have allowed this use of terminology to form the recorded basis of my understanding of his intentions.

When pilots use inappropriate phraseology, ambiguous phrases, or words which do not adequately describe their intentions, controllers are placed in the position of having to *assume* what is meant by their transmissions. Had the pilot in the above example been using the CB code for emergency when he stated "10 something at the airpatch" and had I given him routine handling and not determined exactly what he meant by his transmission, I might have been called to task for not giving this aircraft emergency priority. This may appear to be unfair but, if I did not question what he meant by the use of this terminology, I would have seemed to imply an understanding of that transmission.

Had I been able to discuss this situation with that pilot and explain the fact that his use of inappropriate phraseology was a potential problem, I am sure that he would have been receptive to a friendly discussion and would have discontinued any further use of that phraseology in an aviation environment.

The concept of mutual understanding of key words in air traffic control language along with the ability to be fair and reasonable in our willingness to see others' points of view is critical to the establishment of a safe and effective ATC system. The ability to establish this rapport begins with a pilot's first ground school instructor, continues with his flight instructor's reinforcement (especially through that individual's example), and culminates with the contacts that the pilot has with the ATC system.

I was very lucky when I began to learn to fly, because my first real instructor drove these concepts into my brain with a sledgehammer and practiced them in his everyday flying. He began the phraseology section of his ground school with a little speech that I have always tried to incorporate into any discussions I have with new students. He had a very dry wit and began his discussion as though it was going to be a lecture on physics, then delivered a very common-sense punch line. He drew a picture on the blackboard of a pilot's head with concentric, expanding circles emanating from the mouth on which were written some common ATC phrases. Pointing to the words, he stated that sounds were essentially intelligent noises carried on a vibrating medium commonly referred to as air. Moving the pointer to the head, he continued by saying that this sound was generated somewhere between the ears of the individual making the noise. This, of course, would bring a chuckle from the students, and he would grasp both ends of the pointer and very seriously deliver the punch line. ''My job is to teach you the difference between what is a sound and what is a noise,'' he then turned and pointed again to the head, ''and try to ensure that the medium that generates the sound is somewhat more dense than the medium on which it is carried.''

This man's teaching was peppered with philosophy regarding the need to respect the individual on the other end of the mike and full of challenges to learn more than just the minimum necessary to get by. He always believed that you could accomplish more by discussion than by confrontation, and he insisted that every one of his students spend at least one full day visiting an ATC facility and getting to know the people in that profession. He was probably one of the reasons that I eventually considered air traffic control as a profession.

NO HARM, NO FOUL

In my capacity as an air traffic control supervisor, I occasionally have to talk to pilots or vehicle operators who have done something that is, shall we say, out of the ordinary. I have also been on the other end of the complaint and listened to these individuals when they feel that they were mistreated by an air traffic controller. When possible, I try to keep to the philosophy of ''no harm, no foul.'' No adverse action is taken unless the situation is very grievous or such action is mandated. Most often this discussion is simply a case of having the individual explain their actions or of my having to explain the reasons for why a controller took a particular action. Normally, the discussion will end with an amicable understanding of each other's intentions. I will issue an invitation to come to the facility and plug in with the controllers to see the job from their perspective, and very little else will be done. Every once in a while though, these discussions can lead to some rather interesting statements from the person with whom I am speaking regarding their understanding of ATC phraseology, and I wonder what kind of instruction they had. I feel that I have to do something to ensure that they learn the correct knowledge and procedures, and I wonder what kind of action can be taken to teach them the procedures and not sour them on the ATC system. This book and, in particular, this chapter will hopefully go a long way in that direction.

Fortunately, this ''no harm, no foul'' philosophy appears to be spreading. There is a movement afoot in some elements of the FAA to institute a program whereby a pilot who is identified to have had some misunderstanding of procedures or some gap in the knowledge elements they should have obtained, would be required to attend some type of refresher training to obtain that knowledge. A pilot might even receive the ultimate punishment and be required to spend eight hours plugged into an operating position with an air traffic controller. This program could be designed to be discretionary on the part of the local Flight Standards District Office based on the individual case and the history of the person involved. If we are dealing with a minor problem and/or a first-time offender, this refresher training would be in lieu of any real "punishment," such as suspension or revocation of the pilot's license.

ATC phraseology is designed to be as free from misunderstanding as possible, and there needs to be a mutual understanding of what is meant when certain words or phrases are used. (If I seem to be harping on that one particular idea, I am.) Let's examine a few key words and phrases and give some examples of common pilot usage errors connected with these words and the results of those errors.

RADAR CONTACT

If you ask 100 pilots what is meant by the term *radar contact*, you will likely hear the same basic answer from all of them. It means that the controller is watching you on their radar scope and that you may discontinue mandatory reporting points. With respect to the *AIM* Pilot/Controller Glossary, this definition is accurate, and most pilots and controllers assume that everyone has the same understanding. Unfortunately, there is a second paragraph under this heading in the *AIM* and at least one pilot felt that it substantially expanded the definition of this term.

I had occasion to talk to a pilot who had been radar identified and had then, without permission, flown her aircraft into airspace that required specific authorization to enter. When I explained the situation to her, she replied that she had received permission to enter that airspace from the controller when she first called. I had heard the tape and knew that the controller had not authorized that action, explained that to her, and asked her where she had gotten the idea that she had received permission. She then proceeded to quote me as follows. The term *radar contact* "informs the controller that the aircraft is identified and approval is granted for the aircraft to enter the receiving controller's airspace." She stated that this was exactly what she had done and she did not see that she had done anything wrong.

What she had quoted to me was from the aforementioned second paragraph in the *AIM* and is the meaning of the term *radar contact* when applied to conversations between two air traffic controllers who are in the process of verbally handing the aircraft off between each other. When I informed her of the meaning of that paragraph, she replied, "if that was what they meant, then why didn't they say exactly that." I explained to her the difference in context between "inform an aircraft" and "inform the controller," and she replied that she felt very foolish but that she believed that the FAA should realize that not all pilots majored in English in college and they should be a little more specific in identifying those things that apply only to controllers.

GO AHEAD

The preceding situation was an example of a pilot taking a definition and expanding the interpretation out of literal context. The reverse, or taking a transmission too literally, can be just as bad.

Airport employee vehicle operators are among the hardest-working people that you will find anywhere in aviation. They seem to go out of their way to try to do a good job, and their enthusiasm and willingness to do exactly what you tell them to do, combined with their inexperience in the use of ATC phraseology, will occasionally get them in trouble. I listened to a tape of a runway crossing incident involving a vehicle operator and then later had the opportunity to talk to one of the airport supervisors about the situation. What happened is a classic case of how you hear what you want to hear, and the lesson is applicable to everyone in aviation.

The vehicle driver had pulled out of the maintenance shop and was holding on a vehicle access road short of an active departure runway when he called the ground controller asking for permission to cross the runway enroute to his work assignment. From what the driver said later, he had told the ground controller his vehicle number and location, and stated that he wanted to cross the runway to mow the grass on the other side. Unfortunately, all that you hear on the tape is a loud squeal because at least two transmitters were keyed during most of his transmission. The controller was able to pick out that one of those transmitting was an aircraft on the ramp near the T-hangars, because his transmission lasted longer than that of the vehicle driver. The controller responded to the aircraft and completed his transmissions with the pilot while the driver waited patiently. When the next break between transmissions occurred, the driver called again.

Vehicle Driver: "Ground, this is vehicle 900."

Controller: *Vehicle 900, Ground Control, go ahead.*

You guessed it; that's exactly what the driver did. He assumed that the controller had heard his first transmission and knew what he wanted. When the controller told him to "go ahead," he thought that he had been given permission to cross the runway. I use the example of a vehicle driver to illustrate this point because it is a situation with which I am familiar, but I assure you that pilots have done exactly the same thing. The next few times you overhear a transmission where the words "go ahead" are appropriate, see how many times the controller actually says "go ahead with your transmission."

URGENCY

Not all situations between pilot and controller are covered by a word or phrase that means just one thing or has a limited range of use. It is occasionally necessary to convey a sense of urgency in the transmissions between pilots and controllers, and the level of that urgency is determined by which phrase is used or by the context in which it is used. When controllers are confronted with a situation where a pilot action is required to avoid an imminent situation, we use the word *immediate*. This word, or a series of words that imply the same urgency, is rarely used in ATC communications and, as a result, it has considerable impact when used. Controllers are very careful not to place aircraft into situations where an immediate action is necessary and to not use the word *immediate* unless that is the only action that will work. We try to keep in mind that there is always the possibility of passenger injury or damage to the aircraft when pilots are forced to make radical maneuvers.

Barring any mistakes on the part of the pilot or controller, or unforeseen circumstances, this word is used most frequently in a departure situation. When a controller clears an aircraft for takeoff with another aircraft on final for the same runway, they are basing that clearance on the premise that the departing aircraft will not abnormally delay its departure roll. If the pilot does not begin the departure roll soon enough or if the controller is trying to depart an aircraft in a tight gap, that controller may say, "Cessna 5TA, cleared for an immediate takeoff." In this instance, he is trying to convey to the pilot the importance of rapid compliance with his instruction.

If the controller is conducting an operation that is not quite as time critical, but still requires that the pilot not delay his compliance with instructions, the controller may use a term that is less dramatic but still places an emphasis on the immediacy of that compliance. For example, the controller may say, "Cessna 5TA, taxi without delay, cross Runway 18 Left."

A controller might also place an emphasis on a particular word in a control instruction that contains more than one element so that there is no doubt as to which instruction is to be performed first. As an example, the controller might say, "Cessna 5TA, reduce speed to 210, *then* descend and maintain 5000." In all of these cases the controller is trying to convey a sense of immediacy or establish a certain priority in the mind of the pilot with respect to a control instruction.

All of our ATC procedural handbooks and lists of approved phraseologies contain specific examples of how to communicate these concepts to the pilot, and most controllers are very careful not to convey an unintended sense of urgency or priority to the pilot. Unfortunately, the reverse is not always the case.

Controllers are very sensitive to pilot transmissions which indicate urgency as they are usually related to a situation where a pilot needs assistance. When we hear these types of transmissions, there is almost a shifting of gears to the point where a considerable amount of our attention is directed toward that pilot and aircraft. In these instances it is not unusual for the supervisor and several other controllers to make themselves available to assist the controller involved in the emergency.

Pilots occasionally transmit messages to a controller which imply a sense of urgency when they really did not intend for that to be the case. When this happens, there is a period of wasted time where the major focus of the controller's attention is directed to that aircraft. If the controller then learns that there was no particular difficulty, they are usually faced with a situation where they have to catch up with duties neglected during the supposed emergency. The illustration that we used in Chapter 2, where the pilot asked for vectors to the nearest airport, could easily have been an example of this.

Another situation which grabs our attention is the occasional pilot whose voice patterns are very rapid and give the impression of being "breathless" and panicky. At the outset, these situations are virtually indistinguishable from real emergencies. Until several qualifying transmissions are made, the controller must react as if faced with a real emergency.

We are all trained to give priority to an emergency situation, and each false alert wastes a controller's time and can result in some other important operations being delayed. A couple of these types of pilots on the frequency will make an hour on a control position seem like a roller coaster ride at Six Flags.

In addition to a controller's requirement to analyze what they hear and react to that information, there is also an obligation, in the legal sense of the term, to act correctly on that information. Most of us also feel a strong moral obligation to provide the maximum assistance possible in any given situation.

Let's refer back to the pilot who flew to the closed airport for fuel and practice approaches. Take the following two transmissions and place yourself in the position of the air traffic controller who receives them. Determine what you think the pilot is trying to say and consider how you would react to this information. Remember, this is the first communication you have received after you have told the pilot he cannot land at his primary airport:

Pilot: *Approach, 5TA, we would like vectors to an airport where we can refuel for the trip back home.*

OR

Pilot: *Approach, 5TA, we need vectors to the nearest airport.*

These two transmissions were intended to convey the same information to the controller. Unfortunately, the phraseology used will result in entirely different responses. Try to keep in mind that it is extremely important to use the right words at the right time so that you convey the intended message. Whether you are a student or an instructor, remember that most flight instructors are usually very good about teaching their students what to say in a given situation, but it is only the unusual flight instructor who also teaches their students *how* to say what they say. Again, it is a shared responsibility between the student and the instructor to learn this important aviation element.

EMERGENCY

Throughout our discussion of terms related to a sense of urgency, we have been flirting with the concept of *emergency* operations. This word, or concept, is one of the most important pieces of phraseology that can be exchanged between a pilot and a controller. While it is important that pilots not imply that an emergency exists when it does not, it is considerably more important to admit to an emergency situation when it does exist. Emergencies do happen and, when we know the full extent of a pilot's situation, we can bring some surprisingly efficient help to bear on his behalf.

Unfortunately, pilots will occasionally go out of their way to withhold such information from the controller. Normally, the type of situation which causes this action is one which is not immediately life threatening but does have the potential for serious problems. Often it is related to some failure, on the part of the pilot, to conduct his operation in accordance with proper planning, rules, or procedures. An example of this would be a pilot who was lost or one whose aircraft was very low on fuel. I can only assume that these pilots are afraid that an admission that they require extra assistance will bring down the wrath of the FAA on their shoulders. As a result, they will frequently be vague about their intentions, make requests that are not consistent with an originally stated destination or flight plan, play a guessing game with a controller, or tend to want a controller to make decisions for them about their flight. If you were to listen to recordings of this type of communication you might want to nominate some of these pilots for Oscars. Their acting is better than some professionals I've seen on the wide screen.

Controllers can be accused of having a lot of general characteristics in common, but stupidity is not one of them, and even a controller from Florida can recognize a snow job when he hears one. We will provide the maximum assistance within our capabilities to a pilot who needs it, but we will also not allow ourselves to become part of the cause should you decide to turn your aircraft into scrap metal. Let me give you an example that happened to me.

I was working a radar position that served a group of satellite airports when a pilot called and asked for radar advisories. He advised that he had departed an airport that was approximately 20 miles southwest of the primary airport served by my ATC facility. He further advised that he was about 25 miles southwest of that satellite airport and was enroute to an airport several hundred miles to the south of his location. Since his reported location would have placed him within the airspace served by another ATC facility and since I observed a target in the general location of his reported position, I suggested he contact that ATC facility and provided him with the appropriate frequency.

About 30 minutes later I received a call from a controller at that facility who said that they had been trying to identify an aircraft that was apparently lost and that they were unable to locate him anywhere in their airspace. This controller asked me to look in our airspace to see if I could find an aircraft which should be broadcasting a particular transponder code. I asked the controller for the aircraft ID (the same one that I had spoken to earlier) and entered that call sign together with the transponder code he should be squawking into the ARTS. A few seconds later the ARTS tagged a target that was approximately 30 miles southeast of our radar antenna. I asked the other controller to tell the pilot to "ident" and observed that this target idented almost immediately. Since the aircraft was well beyond the other ATC facility's radar coverage area and was right in the middle of one of our busiest arrival sectors, I suggested that they transfer radio communication to me. (I am probably now on that controller's Christmas card mailing list.)

I pointed the aircraft out to the arrival controller, and when the pilot called I verified that his Mode C altitude readout of 9500 was correct. I informed the pilot of his location and predicament and asked him for information regarding his intentions. He responded that he wanted vectors to an airport where he could "get some lunch" and he wanted to "get lower and out of the weather." At that time, the only weather that we were depicting was a line of thunderstorms on the southwest edge of our radar display. I informed the pilot that he would need to give me a particular destination airport if he wanted vectors and asked him if he would accept vectors out of our arrival sector initially. I also asked him to clarify his request for vectors out of the weather. The pilot responded that he would accept vectors and that he wanted to descend. I offered a suggested heading away from the arrival corridor and advised the pilot that he could descend VFR at pilot's discretion. He responded that he was turning to the heading and he thought he could descend "through the clouds." I hoped that he meant "around the clouds" but only repeated my instruction to remain VFR and then advised the arrival sector controller of what I was doing.

Throughout this operation I was also vectoring several other aircraft, including three for practice approaches, so I had to divide my attention among them and this twin-engine Beechcraft that was cruising around all over the sky. I again asked the Beech pilot for his intentions, and he again wanted me to pick a nearby airport with a restaurant. After several transmissions designed to convince this pilot that I was not going to choose his destination, he finally selected an airport about 70 miles southwest of our facility. I advised him of the weather in this location, read him the current SIGMET which was calling for potentially severe thunderstorms developing in that area, and asked him if he wanted to reconsider. He responded with a question asking what I thought he should do. I considered telling him, thought better of it, and then noticed that he was now descending below a minimum safe altitude and had turned in the direction of a very tall antenna. I issued a safety advisory, suggested a climb and a heading change, and informed him of the location of the obstruction. After we resolved this problem, I considered the situation and decided that I had a moral obligation to the pilot and his passengers (I could hear several voices in the background of his transmissions) that superseded the legal requirements of my air traffic control responsibilities.

I unkeyed the radio and broadcast to the recorder that I was declaring an emergency for this pilot, I stated the time and my operating initials and then returned to the radio. What I was about to do was technically beyond my authority as a controller, so this declaration of emergency was my way of stating my reasons for taking this action. I did not want to alarm the pilot with a statement like this over the air, and I felt that it was necessary to do what we call CYA (cover your aft-section) before I stuck my neck out. I then spoke to the pilot as follows.

> N123TA, you have been in the air for over an hour, your current location is 20 miles east of the airport from which you departed, your requested destination will take you into an area of severe weather, and you have made zero progress toward that destination. I strongly suggest that you return to ________ airport, land, and reconsider your options.

Perhaps the pilot's spouse was aboard the aircraft (they have a way of convincing husbands of the wisdom of logic) because, to my surprise, the pilot seemed to be relieved that someone had made his decision for him. He thanked me and requested vectors to that airport. After several additional vectors needed to steer this aircraft, he sighted the satellite airport, left my frequency and, I hope, landed safely.

Except in the case of a clear emergency where a controller's experience can mean the difference between life and death, controllers do not like to have to make suggestions to pilots or make decisions for them on non-ATC-related matters. Inevitably in these situations, the controllers are not aware of the skill level of the pilot, frequently are in the dark about the circumstances that lead pilots into these predicaments, and often have to deal with pilots who are unwilling or unable to be totally frank and honest with us. For us to unilaterally take charge of the situation and begin to make decisions normally reserved for the pilot in command is an open invitation for criticism of our ideas or techniques should that operation result in some type of problem. We will not normally assume that role unless, in our judgement, the lives of the aircraft occupants are in jeopardy.

If the tendency to involve controllers in the pilot's decision-making process was limited to just a few pilots, we could probably identify those individuals and correct the problem. This is unfortunately not the case. Pilots also try to play the CYA game, and to my chagrin and anger, some flight instructors actually teach student pilots how to get out of trouble without admitting that they need help. You would be surprised at the number of requests that we get for practice ADF (automatic direction finder) steers or requests for vectors to an airport which require pilots to make a turn of greater than 90 degrees to establish an on-course heading to that airport. One would expect that a pilot who is not lost would at least be heading in the general direction of his destination airport.

Much like the rest of society, controllers have to make split-second decisions amid noise, tension, and incomplete data, and then contend with a legal system that has the time and the hindsight to microanalyze every statement, inflection, and possible reaction. We understand that pilots are reluctant to confess to a problem when they have had a hand in causing that problem, but when controllers encounter a pilot who is trying to circumvent the system while laying the responsibility for their decisions on us, we become very protective of our own aft section and will decline to play the game.

If you are experiencing difficulty, explain that difficulty to us and we will assist you to the maximum extent possible. If this situation does not constitute an emergency, there is every likelihood that we will simply provide whatever assistance is required and then let it go at that. Virtually all controllers have to deal with students in a learning environment on a daily basis. We assist them routinely, sometimes extensively, and then just go about our business. A situation has to be a relatively serious violation of procedures or safety concepts before we ask the aviation inspectors to investigate the circumstances. Remember, if your situation is (or could develop into) an emergency, you should be smart enough to want every possible advantage directed your way.

There are two very important things that all pilots need to remember when dealing with the ATC system in a real or potential emergency. First, the pilot is the final authority in the aircraft and it is not likely that you will ever find a controller who will refuse to help a pilot who asks for emergency assistance. Secondly, the FAA is in the business of making the air traffic system safe, orderly, and expeditious. Ours is not a Gestapo organization that exhausts its resources in finding and punishing pilots who find themselves in difficult situations. In most cases, a statement from the pilot that he or she felt that the circumstances indicated a need for a declaration of emergency will be adequate justification for that action and that will be the end of the discussion. It is usually only the rare, blatant cases of pilot error that go beyond the facility level to an investigatory organization within the FAA.

ADDITIONAL SERVICES

Let's move on to another category of phrases that cause some heartburn to both the pilot and the controller. Several of the controller actions that I described in the above situations fall under the category of "Additional Services." From the standpoint of the average pilot, this is one of the most misunderstood concepts in aviation and is the source of considerable erroneous pilot expectations and frustrations when they do not receive services they feel they have a right to expect. Allow me to quote the definition of Additional Services from the *AIM* and then we can discuss some of these erroneous expectations.

> ADDITIONAL SERVICES—Advisory information provided by ATC which includes but is not limited to the following:
>
> 1. Traffic advisories.
> 2. Vectors, when requested by the pilot, to assist aircraft receiving traffic advisories to avoid observed traffic.
> 3. Altitude deviation information of 300 feet or more from an assigned altitude as observed on a verified (reading correctly) automatic altitude readout (Mode C).
> 4. Advisories that traffic is no longer a factor.
> 5. Weather and chaff information.
> 6. Weather assistance.
> 7. Bird activity information.
> 8. Holding pattern surveillance.
>
> Additional services are provided to the extent possible contingent only upon the controller's capability to fit them into the performance of higher priority duties and on the basis of limitations of the radar, volume of traffic, frequency congestion, and controller workload. The controller has complete discretion for determining if he is able to provide or continue to provide a service in a particular case. The controller's reason not to provide or continue to provide a service in a particular case is not subject to question by the pilot and need not be made known to him. (See Traffic Advisories, *AIM*.)

Even given the above definition, some pilots erroneously believe that when the controller says the magic words, "Radar Contact," they can relax and let the controller take care of everything. They are shocked when a controller issues a heading that will take them through an area of weather that they can see right out of their window, or when another aircraft goes swooping past them that was not reported as traffic by the controller. These pilots don't understand the concepts and limitations of Additional Services nor do they understand the equipment limitations that affect those services. I'll talk about equipment limitations in general in the next few paragraphs and in greater detail in Chapter 4 where I'll give some specific examples of how this equipment works, but for now let's talk about those "additional services."

First, let me state that all of the services described above are mandatory on the part of the controller. We cannot simply decide that we will not perform any or all of these services. The discretionary aspect of these services means that we will first perform higher-priority duties, such as the separation or the sequencing of aircraft. If the sheer volume of these duties does not allow for the application of additional services, they can be delayed until time and/or resources exist to perform them.

If a controller agrees to identify and provide service to a VFR aircraft, he or she may, depending on the type of services being provided, have also accepted an additional separation requirement. A controller who is unable to accept this additional separation responsibility may decline to identify that aircraft even though the controller may be able to provide some or most of the other services that go with working that aircraft. To us, it's a case of: when you buy into the front end of a mule, the back end comes with it. When we say yes to part of the job, we accept responsibility for all of the job, and we don't want to do a half-____ed job.

Additionally, even though a controller may have identified an aircraft and begun to provide additional services to that aircraft, the application of these services may tend to fade in and out depending upon the workload at the moment. When workload is light we will probably issue weather advisories to you in depth by soliciting PIREPs, contacting the FSS specialist, and conducting an analysis of everything that we see around you. When we are very busy separating and sequencing traffic, we do not have time for this entire analytical process and we will provide weather/traffic advisories based on what we have, combined with our knowledge and experience. We like to think that we always provide the best information available; it's just that you sometimes get the condensed version.

Pilots sometimes have difficulty understanding that equipment limitations also enter significantly into the factor of how much information we can provide to them and when it can be provided. A controller will issue traffic to a pilot, and then several minutes later another aircraft may pass dangerously close to that pilot without as much as a hint from the controller that this aircraft was there. The pilot has been frightened and, just like everyone else when they are afraid, he will lash out angrily at the only one available, in this case the controller. The controller may have had his/her attention distracted elsewhere because of higher priority operations, or that target simply may not have appeared on his/her radar display. One possible cause of this last condition could be loss of radar tracking due to the path of the aircraft. This is only one of the limitations of the airport surveillance radar system which we will discuss in Chapter 4.

Similar limitations apply to our ability to provide pilots with weather information. We often receive requests from pilots for vectors through an area of weather, and these requests put us in a particular bind. Controllers want to assist pilots to the best of our ability, but we also do not want to place ourselves in a position of providing this additional service and then appear to have misled someone into believing that all is well. If we acquiesce to these requests, most pilots mistakenly believe that we are going to vector them away from all of the bad weather that exists in their environment. When we are unable to do this, the pilots get mad at the controllers. These pilots think that the controllers are not paying attention to their situation or are giving them some sort of run around to avoid doing their job. Again, this is due to a fundamental misunderstanding pilots have of the controllers' equipment, its limitations, and the controllers' ability to accept responsibility for actions within those limitations.

In today's legal system, controllers are finding that they are increasingly being included as defendants in lawsuits and, in some cases, being held accountable for actions or requirements which had previously been considered the sole responsibility of the pilot. As a result, I, for one, am generally more cautious in handling tricky situations, and I have to consider the possible implications of a pilot request for additional services that could involve me in a potential problem. Here's how I would handle a typical pilot request in light of this:

Pilot: *XYZ Approach, this is N123TA, VFR from point A to point B. We would like VFR advisories and vectors around the weather.*

Controller: N123TA, XYZ approach, are you equipped with weather radar?

Pilot: *Negative, approach. That's why we need vectors around the weather.*

Controller: N3TA, be advised that our radar systems are not equipped to depict all of the weather that you might encounter. I can offer you suggested headings around the areas of precipitation that I see but there may be other weather areas that are not depicted on my radar display. Can you accept that?

This pilot had begun his transmissions with me by stating that he wanted me to keep him out of the weather. The implications of that statement suggest that he expected to be vectored away from all potentially dangerous weather situations. I was willing to provide the pilot with the basic vectoring and weather advisory services that were within the capabilities of my equipment, but I was unable to guarantee that I could "keep him out of the weather." Since I did not want him to proceed on the assumption that I was going to do that, I had to establish, and have him acknowledge an understanding of, the limitations of what services I could reasonably offer before I identified the aircraft.

Had I just identified this aircraft and begun issuing vectors without qualifying what I was doing, the pilot could have reasonably made the assumption that I was doing exactly what he had asked. This is a dangerous assumption which would have resulted in a very relaxed vigilance on the part of the pilot and the possibility that he, and I, would not have seen a severe weather situation. You might suggest that no pilot in his right mind would place his life so totally in someone else's hands, but when you consider the fact that pilots operating in close proximity to each other in IFR conditions are doing much the same thing, this is simply an extension of that trust.

Part of the problem in providing additional services of this type is that controllers, and the ATC system in general, have been so successful in that endeavor in the past that pilots have come to expect much more than a controller can reasonably deliver. While we value that trust, we do not want it given blindly and we want you to know the limitations of what we can do.

CLEARANCES

Virtually any service that we provide for a pilot, including those that we have been discussing up to this point, require some type of air traffic control action to activate. This action is usually an ATC *clearance*. Defined as an authorization for aircraft to proceed under the conditions specified by the air traffic control unit, a clearance is some type of a positive statement by an ATC specialist that usually, but not always, contains some form of the word *clear*. This word, or other action statement, and the concept that they imply, is frequently misunderstood by pilots, especially those who do not normally deal with the more complex aspects of the ATC system.

These pilot mistakes seem to come in all forms and often are at opposite ends of the possible definitions that could be applied to the word. Some pilots consider any transmission from a controller to be a form of a clearance. Others know that they have to have a clearance from a controller but they seem to believe that, once they have that clearance, they are authorized to do anything that they want. These pilots seem to think that a controller has to specifically tell them if he or she does not want them to do a particular action connected with that clearance. Still others will do things that make you wonder if they have the slightest notion of what a clearance involves. Let me give you some examples of things I have seen over the years that illustrate these problems.

I was working in the tower one afternoon when I received a call from a pilot who advised me that he was on a cross-country flight to a small field about 20 miles north of the airport where I was working. The transmission that the pilot made was given in a format that made it seem informational in nature. (We get a lot of transmissions from pilots who just want us to know that they are there.) I looked at my radar display and noted that the track I observed on the radar display which corresponded to his reported position indicated that his flight path would take him well clear of the airport traffic area (ATA). I asked the pilot to ident just to verify in my own mind that the track I was watching belonged to the pilot on the radio, rogered his transmission, and gave him the current altimeter. The pilot thanked me and said, "Good day."

As I watched the radar display I saw the target's track make an almost immediate turn and start in a direction that would now take it directly through the ATA. I asked the pilot if he intended to fly through the ATA, and he replied that he thought he had my permission to do so and he seemed to be honestly surprised at the question. I explained to him that I had only responded to his call on the frequency and that if he wanted to fly through the ATA without landing at an airport within the ATA he had to specifically ask for permission to do so. I then advised him that there was no problem and that he was now cleared through the ATA. I asked him to fly directly over the airport so that he could avoid traffic that would shortly be on final. The pilot was very apologetic, and he thanked me for the information.

This pilot should have been taught this information by his flight instructor, but at least he was operating from a position of ignorance and not trying to bend a rule. Some pilots allow themselves a very liberal interpretation of these rules and requirements, and in Chapter 5 I'll show you an example of a similar situation with a pilot who twisted the rules like a pretzel.

SURPRISE, SURPRISE, SURPRISE

One thing that drives controllers up the proverbial wall is a situation where they clear a pilot for one type of operation, establish a sequence based on what they expect that pilot to do, and then have to scramble when the pilot does something totally unexpected.

One day I was working a very busy satellite position during marginal weather and having to conduct a VOR circling approach to a busy airport. Most controllers do not enjoy this type of operation when we are busy because even a straight-in VOR approach tends to increase the interval between aircraft. Due to the higher approach minimums, we see more missed approaches here than on precision approaches and frequently have to work the same aircraft more than once. Pilots also like to operate at a slower speed on a non-precision approach and are frequently less practiced on these types of approaches than on an ILS. The VOR approach that I was using at that time was based on a VOR located about 10 miles east of the airport, and I was vectoring to a point about 13 to 17 miles east to set up a straight-in VOR approach (SEE DIAGRAM).

Approach plates for VOR, NDB, and other types of non-precision approaches frequently contain information which indicates any restrictions to that particular approach. These restrictions might either be something that is required or something that is not authorized. For example, if a straight-in approach is not authorized, it is so noted. On the other hand, if a procedure turn is not required from a straight-in vector, that is also noted.

We were about one third of the way into what was going to be a very busy arrival push, and I was vectoring several aircraft on the downwind on either side of the final approach course to keep the final from extending too far from the

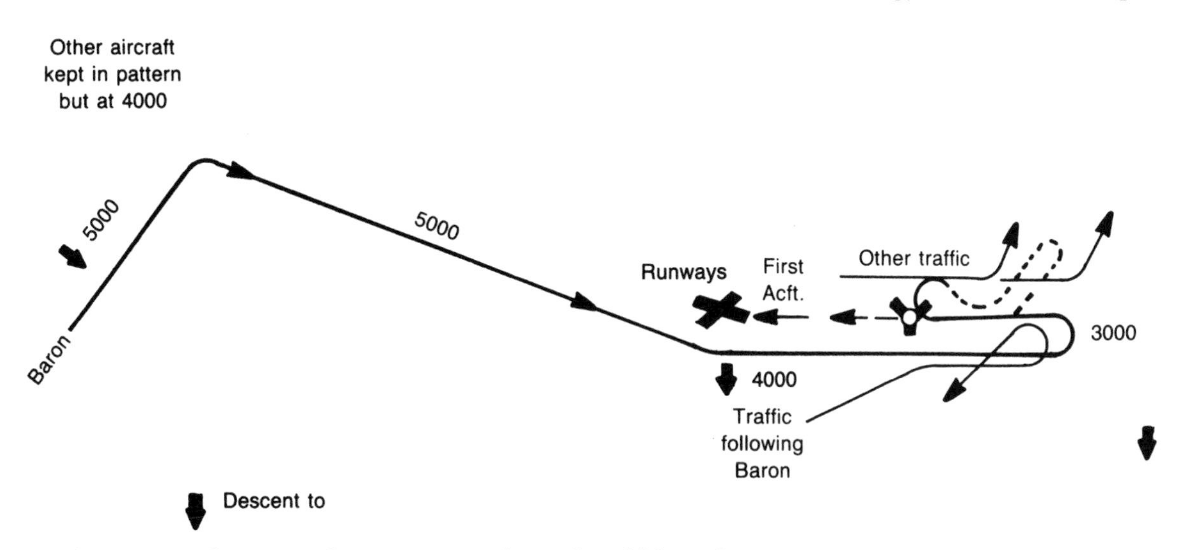

Making a procedure turn where it is not authorized could have disastrous consequences. At the very least it will surprise the approach controller. Here, other traffic had to be scattered when the Baron decided to pull this surprise.

airport. The aircraft that I had projected as next in the sequence was a Beech Baron whose pilot had been complaining about the sequence, the fact that he had been slowed early, and the fact that he was still 1000 feet above the VOR crossing altitude. When I turned the aircraft toward the VOR he was descended from 4000 to 3000 feet (there was no longer any aircraft operating below him), was operating at 160 kts (which was 30 kts faster than the Cessna 182 that was in front of him) and was on a 17-mile final which placed him 7 miles from the VOR (plenty of time to lose 1000 feet) and was slightly over 8 miles behind his traffic. This sequence would have placed the Baron on about a 5-mile final when the Cessna 182 landed. I issued the pilot the following instruction. ''Maintain 3000 until over the VOR, then cleared for the VOR approach to Runway 27, circle to land Runway 2.''

The position where I had placed the aircraft was virtually on the extended centerline of the VOR final approach course on the side of the VOR that was away from the airport. Because of the positioning I had given this aircraft, the fact that he had been on the frequency during the time that I had cleared several other aircraft for the approach, and the fact that a straight-in approach was depicted on the approach plate along with ''NoPT'' (no procedure turn) notation, I fully expected this aircraft to conduct a straight-in VOR approach.

When the pilot read back his clearance my job was essentially finished except for monitoring the aircraft's progress on the final approach course, and I instructed the pilot to contact the tower over the VOR. I then turned the majority of my attention to positioning the next aircraft into the sequence.

By the time that the Baron reached the VOR, I had already descended the lead aircraft on both sides of the downwind and turned the one on the south side of the final approach course to a northwest heading which would intercept the extended centerline. To this day I still consider the fact that I was watching the Baron's target when it crossed the VOR to be one of the most fortunate circumstances in my career. I noticed that the Baron appeared to be turning to the north and I knew that, even if he returned immediately to the final approach course, my sequence for the next aircraft had been destroyed. I immediately turned that aircraft toward the south anticipating that I would eventually give it a 360-degree turn right back to the final. I attempted to call the Baron but the pilot did not reply, which I assumed to mean that he had switched to the tower.

While I was attempting to contact the tower it was rapidly becoming apparent that the Baron was going to make a full procedure turn before completing his approach, and I needed to either stop him from doing that or scatter several aircraft out of his way. The headings he would take to accomplish this maneuver were going to place him in conflict with at least two more aircraft, and I decided to get the other aircraft out of the way before the situation deteriorated beyond the point where safety could be ensured. By the time I cleared enough airspace for the Baron to conduct his turns I had moved four aircraft out of the sequence and my pattern looked like a Keystone Cops movie set.

When members of our facility discussed this situation with the pilot a short time later, the attitude he took seemed to fall along the lines of the theory that a good offense is the best defense. He felt that since he had been cleared for the approach he could conduct the approach any way he wanted. He stated that if we had wanted to place any restrictions on his ability to conduct that approach, it was our (the controller's) responsibility to do so when we cleared him. He made several comments indicating the level of esteem he held for controllers and generally challenged us to do anything about the situation. I don't really know what the FAA did about this incident, and I'm not absolutely sure that we could have made anything stick if we tried. I must admit that I would like to meet this pilot, preferably in a country where dueling is still allowed.

VERY SPECIAL VFR

As bad as the preceding incident may seem, perhaps the craziest thing I have ever seen regarding the issuance of a clearance had to do with a pilot trying to obtain a Special VFR clearance.

The facility where I was working at the time had two fairly active flight training schools, a moderate amount of itinerant traffic, and only a few air carrier operations each day. The tower at this field was located in the middle of the airport and the ramp and major portions of the main runway were not visible from the tower cab when the visibility was less than 1½ miles. As a rule, Sunday mornings at this facility were a particularly quiet traffic time, and the only normal operations were one air carrier arrival at 7:30 A.M. and four to six pilots who, when the weather was VFR, always flew to a nearby airport for breakfast. Because of this light traffic, our Sunday staffing amounted to one person in the tower and one person in the radar room until 8:00 A.M. On this particular Sunday, the field was IFR with a ceiling of about 700 broken and visibility between 1 and 1¼ miles with light rain. The place was dead. At about 7:10 A.M. an aircraft called ground control:

Pilot: *XYZ ground, Cessna N123TA, T-hangars, taxi for takeoff.*

Controller: N123TA, XYZ ground, the field is IFR with ceiling 700 broken and 1 mile visibility with light rain, advise your intentions.

Pilot: *Roger we're ready to taxi.*

Controller: N3TA, I say again the field is IFR, have you filed an IFR flight plan?

Pilot: *Negative, I don't have an IFR rating, we're ready to taxi.*

Controller: Cessna 3TA, the field is IFR and I am unable to authorize a VFR departure at this time, advise your intentions.

Pilot: *Are you telling me that I can't take off?*

Controller: Sir, unless you are operating on some type of flight plan other than VFR, I cannot authorize a departure when the weather is below VFR minimums.

Pilot: *Ok, I guess we'll taxi back to the hangars.*

I was mentally shaking my head at this guy and thanking my lucky stars when our old pal Murphy decided that the timing was too perfect to pass up. About 30 seconds later I received a call from another aircraft:

Pilot: *XYZ ground, this is Piper N345TA, GA ramp with Information Alpha, ready to taxi, requesting a Special VFR departure out of the control zone to the north, and we will be ready for an immediate departure.*

Controller: Piper N345TA, taxi to Runway 27, and advise ready to copy your clearance.

Pilot: *Ready to copy and we can switch to the local control frequency if you like, we're in a bit of a hurry.*

This was obviously a locally based pilot who knew that I was operating and transmitting on both positions and would have to change him to the local control frequency before I issued a departure clearance:

Controller: Cessna 5TA, you can read back your clearance on the local frequency if you wish, and you are cleared out of the control zone to the north to maintain Special VFR conditions at or below 2500 MSL, departure control frequency will be 117.2, squawk 0301.

Pilot: *Cessna 5TA copied the clearance and we are ready to go.*

The pilot was transmitting on the local control frequency so I advised the radar controller of the proposed operation and obtained a release for the aircraft. I then unkeyed the ground control transmitter so that this frequency would be usable while I was talking to the pilot on local control, cleared the aircraft for takeoff, and asked him to advise me when he started his turn northbound. When he did that, I instructed him to contact departure.

Shortly after I completed this operation the radar controller called up and advised me that an IFR aircraft was inbound for the ILS approach and would be landing within five minutes. Just as I finished talking to the radar controller, our old friend N3TA called again:

Pilot: *XYZ ground, this is Cessna 3TA. We would like one of those special things, and we're ready to taxi.*

To a controller, that kind of transmission is like an invitation to walk across a mine field. We cannot refuse a legitimate, legal pilot request, but when a situation like this presents itself, we will go into a CYA mode and virtually require a pilot to use the exact phraseology necessary to obtain any clearance.

Controller: Cessna 3TA, the field is still IFR, and VFR operations are still not authorized, advise your intentions.

Pilot: *We want one of those special clearances to the north, and we are ready to taxi.*

Controller: Sir, this airport is IFR and if you wish an authorization from me to depart, you will have to tell me exactly what it is that you want to do.

There was total silence on the frequency for several seconds (I assume that the pilot was consulting with the others in the aircraft) and then the pilot replied:

Pilot: *Cessna 3TA would like a Special VFR clearance to depart to the west.*

Well, the man had said the magic words, so I instructed him to taxi to the runway and asked him to advise when he was ready to copy his clearance. When the Cessna advised ready to copy I issued the Special VFR clearance and required the pilot read back to ensure that he understood exactly what was said. I then instructed the pilot to change to the local control frequency and advise when ready. Right about this time the inbound IFR aircraft called over the outer marker and I cleared that pilot to land. About two minutes later the IFR aircraft landed and, at this point, Cessna 3TA had not yet called ready to depart. Just then the radar controller called up to me and asked if I knew anything about the target he was depicting about three miles west of the airport. This was one of those situations where you get a cold chill up your spine because I was absolutely sure who that target was going to be.

We used an information transfer procedure between the tower and radar positions which involved writing the flight plan information on a paper strip which was placed in a plastic strip holder. If the Cessna had called ready to depart, I would have dropped this holder down a drop tube to the radar position and called for an authorization to release a Special VFR aircraft into the airspace around the airport. Since the Cessna had not called ready for departure, the radar controller would not know anything about this traffic. I called the Cessna on the local frequency and asked him his position. He replied that he was about four miles west of the airport. I asked him why he had departed the airport without a departure clearance and he responded, "You cleared me out of the control zone to the west," and stated that he had done exactly what I had told him to do.

This aircraft had departed an IFR airport while another aircraft was on final, probably inside the outer marker, and the pilot obviously did not understand the concept of how a Special VFR clearance differed from a departure clearance. Because of the situation and the possibility that less-than-standard separation may have existed between the Cessna and the IFR arrival, we instructed the pilot to contact the facility during business hours and began the process of reporting the incident. The subsequent investigation by our own facility and the regional General Aviation District Office (GADO) personnel concluded that, given the timing of the transmissions, there was no way to determine whether loss of separation had actually occurred. The inspector's interview with the pilot ascertained that he was not familiar with Special VFR procedures, thought that a clearance out of the control zone was the same as a clearance for takeoff, and assumed that the departure frequency assignment was for Stage III (radar sequencing and advisories) service which he never used.

It was determined that I had not acted improperly, although it was suggested that I could have specifically instructed the pilot to hold short of the runway which might have caused him to question whether he was allowed to depart. This type of transmission was not actually required on my part, but the idea of restating the obvious and requiring a pilot to verify compliance with FARs with which he should be familiar has some validity. I now do that in a situation where the presence of that knowledge might be called into question by the circumstances. I do not know exactly what action was taken against this pilot, but I suspect that the GADO inspector wanted to have some serious discussions with him.

CRUISING FOR A BRUISING

Each clearance is a designed procedure which allows a pilot to conduct an operation in an orderly manner. Some are designed to identify preestablished routes, altitudes, and procedures to follow in conjunction with precision operations, while others are designed to allow the pilot the maximum flexibility to conduct a procedure within the boundaries of common sense and safety. An example of the latter case, and perhaps the most misunderstood clearance of all, is the *cruise clearance*.

I once had a pilot tell me that the cruise clearance was a procedure that had been brought about as a result of intense lobbying pressure brought to bear on the FAA by a general aviation pilots' organization. He stated that pilots were being forced to cheat on their altitude in the vicinity of airports that had no published instrument approach and that this procedure was established to legalize the pilot's actions. He also praised the FAA for having the courage to recognize that pilots were going to do what they wanted and for developing rules that supported those actions. Consider the implications of that logic for a while and then sit down at a control position with that pilot on the frequency.

A cruise clearance is, and again I will be quoting in context from the *AIM* Pilot/Controller Glossary, simply a clearance which authorizes a pilot "to conduct flight at any altitude from the minimum IFR altitude up to and including the altitude specified in the clearance." This minimum IFR altitude may or may not be lower than the minimum altitude at which the controller can vector an aircraft in a particular area but, in either case, it allows a pilot the flexibility to maneuver into a position where VFR flight can be conducted. When used in conjunction with an airport clearance limit at locations that are within/below/outside controlled airspace and without a standard/special instrument procedure, this clearance allows a pilot "to proceed to the destination airport, descend, and land in accordance with applicable FARs governing VFR flight operations." More importantly, and the cause of a lot of pilot misunderstanding, is a list of what this procedure does *not* authorize. It does not authorize a pilot to descend under IFR conditions below the applicable minimum IFR altitudes, nor does it imply that ATC is exercising control over aircraft in uncontrolled airspace. Controllers will work with a pilot to provide them a clearance which gives them the best possible chance of conducting their operation, but we will not condone, authorize, or imply that we are letting them cheat.

The issuance of a clearance is, for all practical purposes, a pilot's legal authorization to conduct the operation in question. As with any other type of legal authorization, there are specific steps that are expected to be followed and specific limits to the scope of that authorization. If there is ever a question in your mind as to what those limitations are with respect to an aviation procedure, you should learn the answer before you climb into the aircraft or, at the very least, before you agree to that procedure.

HURRY UP AND WAIT

Another term that I think requires discussion in this book and one that has been used in aviation circles more frequently in recent years is the concept of *delays* as it applies to air traffic control. There are some who suggest that the FAA is actually in the business of air traffic delays as opposed to air traffic control. For those of us who are on the front lines of the ATC system it is very frustrating having to deal with individuals, particularly pilots, and/or groups, including the media, who don't really understand what constitutes a delay.

The concept of what is and what is not a delay depends partly on the point of view of the individual being asked. A lot of pilots who complain about delays seem to think that any time they cannot do exactly what they want, when, and in the exact order that they want to do it, they have been delayed. Controllers, on the other hand, tend to look at the long view in terms of which operation will result in the fastest, most efficient operation which provides the least delay for all aircraft involved. They feel that they have caused a delay for an aircraft when too much time or airspace is wasted in what otherwise would have been that aircraft's normal sequence. When we blow a call and cause a delay, we also increase our own workload. I can assure you that we do not do this intentionally and, if a controller does it very often, he or she will not advance very far or last very long in this business.

We could get into a long-winded discussion about planned ground delay programs, enroute spacing procedures, etc., but this book is about common misunderstandings that exist between pilots and controllers on individual, day-to-day operations, so let's talk about something down on our level. Perhaps here we can come to a mutual understanding of why controllers use the procedures that they do.

For example, let's assume that a controller is faced with establishing an IFR sequence between two Cessna 172s, a PA28, and a Lear 25. We will further assume several other things. First, all four of these aircraft will be in position for vectors to the same ILS final approach course that has outer marker 5 miles from the runway (SEE FIGURE). Second, all of these aircraft will be in position to execute that approach within two minutes of each other. Finally, if time over the arrival fix were the only consideration, the two C172s would be first, the Lear would be the third aircraft in line, and the PA28 would be last.

In IFR conditions, the first aircraft would have to be turned on the final at a point about 7 miles from the landing threshold. The second aircraft, also a C172,

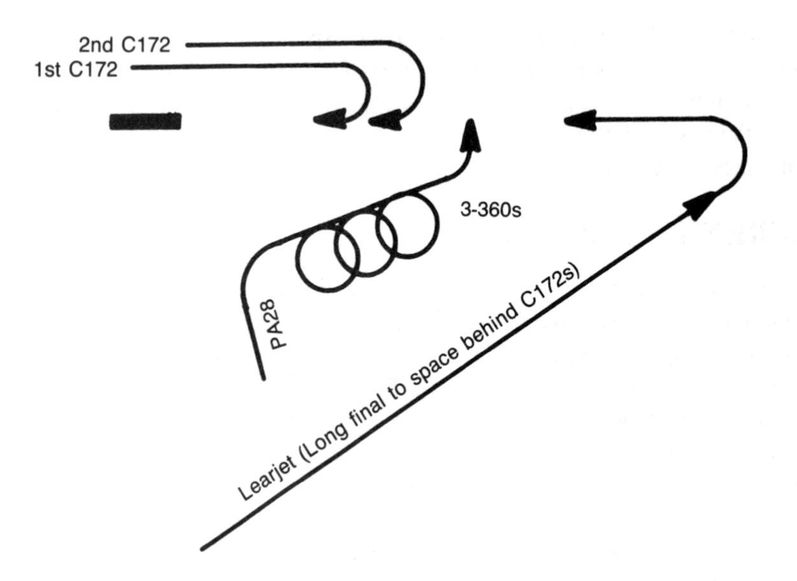

In air traffic control, a first-come, first-served philosophy is often too inefficient for the traffic mix and volume. Here the result is substantial delay to both the Learjet and the PA28.

could be established about 3.5 to 4 miles behind the first, or about 11 miles from the threshold. Both of these aircraft have an approach speed of about 80 to 90 knots so we will use an 80-knot figure to ensure that the minimum allowable separation of 3 miles will never be lost. The Learjet, with an approach speed of 160 knots, or twice that of the Cessnas, will have to be at least twice as far from the threshold as the second Cessna when it is turned on to the final. Actually it will have to be farther than that because we must still have the minimum of 3 miles radar separation between them when the Cessna crosses the landing threshold. This will put the Learjet on at least a 25-mile final (26 if the controller has any sense) when it is cleared for the approach. The PA28 will now have to begin its final approach leg from a point 28 to 29 miles from the runway just to ensure an initial minimum radar separation behind the Learjet. In reality this Piper would be given several 360-degree turns to keep it in close and then turned on to an 8- to 9-mile final after the Lear passed.

For the sake of argument we can reverse the sequence and place the Learjet number one on a 7-mile final (SEE FIGURE). Since this is the fastest aircraft, we can now place the two C172s on 10- and 13–14-mile finals respectively and be assured that the spacing will remain constant. When compared to the first scenario, this reduces this interval between the first and third aircraft by about 15 miles or almost 6 minutes of flying time for the Lear. In this sequence we should only have to give our Piper PA28 one 360-degree turn and then vector that aircraft

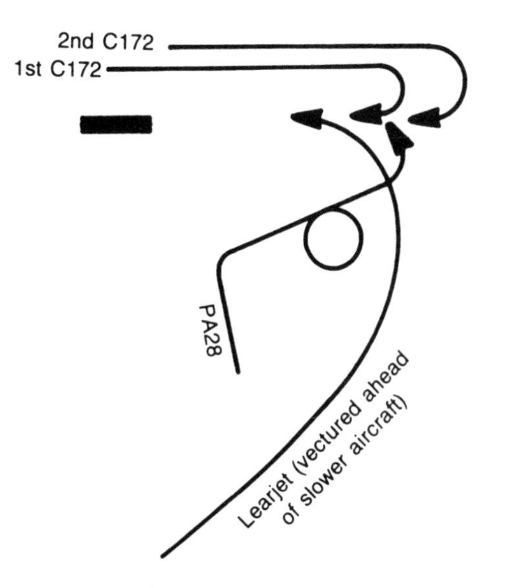

By vectoring the Learjet ahead of the other aircraft, the 172s are delayed slightly, but the Learjet and the PA28 benefit greatly.

to a 10-mile final. Now let's examine the numbers to see which sequence resulted in the least overall delay.

In the second scenario, the two Cessna 172 pilots could, from their point of view, have legitimately said that they had received a delay. They had to fly an extra 3 miles each or a combined total of about four minutes of flying time at 10 gallons of fuel per hour. But from the other point of view, the Piper pilot only had to make one 360-degree turn instead of the three such turns that went with the first scenario. At a standard rate turn of 3 degrees per second, that works out to a savings of 4 minutes of flying time for the PA28 at 10 gallons per hour. The Learjet would have had to fly an additional 36 miles (the difference between a 7-mile final and a 25-mile final multiplied by two to include the downwind) in the first scenario as compared to the second, all at 160 knots which is at or near its worst ratio of speed to fuel burn. Given the total picture, which scenario would you consider to have been the one with the most delays?

What I have just described is a typical ATC sequencing situation that controllers face hundreds of times each week. Don't get the idea that we routinely place the bigger, faster aircraft first as a matter of policy. Most sequencing is done among like types, and other factors enter into every sequencing decision. For example, given the correct, advantageous intercept angles, I would not hesitate to sequence a C172 ahead of a twin-engine aircraft if those aircraft were able to maintain relatively similar speeds (within 30 knots) and all other factors were equal. Additionally, the circumstances of a particular situation often make the sequencing requirement academic. If the slower aircraft happened to be on a straight-in final from a long distance out, we would vector aircraft in front of him until we ran out of room to do so. At that point this aircraft would automatically become next in line and we would begin sequencing behind him. In short, most controllers are going to do whatever is necessary to place the maximum number of aircraft on the ground (and out of our area of responsibility) in the shortest and safest period of time.

Even when a controller says, ''expect no delays,'' he or she means that, given the routine requirements of separation, you can expect to conduct your operation in its most normal, logical, safe sequence. When two or more aircraft want to do essentially the same thing at the same time, only one of them is going to be first. The pilot of every other aircraft is, at least in his/her mind, being delayed. In terms of the logical sequencing of aircraft, this is a no-win situation for an air traffic controller. We are always going to set up the interval which allows a pairing of the concept of first-come, first-served together with the requirement to obtain the maximum utilization of time and airspace. We hope that this will then realize the best, fairest sequence for all concerned.

CONTROLLERS DO CARE

We have covered a lot of terms in this chapter and could probably continue this discussion with an analysis of the meanings of any number of other words and concepts. But the main point that I am trying to make in this chapter, and the previous chapter, is that the language of pilot/controller communications is so unique, so critically important, that it must not be misunderstood or taken lightly. Most of the examples that I have used to illustrate these concepts can be considered amusing in retrospect and are often the source of war stories told in the break room by controllers. We are shaking our heads and laughing about the situation and the crazy pilots who got themselves into those predicaments. But I assure you, at the time these situations were developing, the controllers were not in a laughing mood. Pilots occasionally complain to us about controllers who snap at them over the frequency. They insist that we do something about this discourteous behavior. While I agree that this type of behavior is uncalled for, and assure you that we do in fact do something about it, perhaps I can put these situations into context and explain why we sometimes get short.

Those of you who are parents can probably relate to a situation where your child, about whom you care very much, does something which requires that you rescue them from near disaster. Your immediate reaction is based on an adrenalin surge, and after you have done whatever it takes to resolve the problem, you are left with that residual fear that you express as anger. While you are not particularly angry at the child, you are expressing anger to vent your own frustrations and fear at the fact that this child unknowingly placed himself in danger.

Controllers often operate near what we call the ragged edge, where the difference between a good operation and a possible incident, accident, or systems error can be described by one pilot misunderstanding or one blown sequence. When we make a mistake, we are very critical of ourselves, and when you make a mistake, we sometimes react like a parent does with a child.

When a controller issues an important control instruction to a pilot and receives nothing but dead air in return, the tension starts to build. The controller is thinking, "Come on captain, if I had time to do this twice, I wouldn't have done it in the first place." Two or three such failures from the same pilot are likely to elicit a wake-up call. When you are operating in a busy (not necessarily large) airspace area and blow an altitude assignment, there is a strong probability that someone else is operating near that altitude. The reaction from the controller is going to be swift first, tactful later. Don't get the idea that I am trying to condone this type of response, I'm not. Rather, I am just trying to explain why it sometimes happens. It is almost like we are saying, "How dare you place yourself in danger, and my career in jeopardy, on my frequency and in my area of responsibility."

This is one of the few professions where people's lives, livelihood, and safety depend on each other's understanding of the spoken word. I consider it ludicrous that some people would climb into an aircraft and launch themselves into the sky without a total understanding of the meanings and limitations of each and every one of those words. I would hope that you are using this material, along with whatever other material is available, to educate yourself beyond the point where that can be said of you.

4

ATC Equipment

BEYOND THE OCCASIONAL CLOSE ENCOUNTER WITH A RADIO ANTENNA OR A misplaced landing on top of a localizer shack (yes, one pilot actually did that), pilots rarely cause problems for ATC equipment. The most frequent difficulties that we have with pilots in this area, and the reason that I include a discussion of equipment in this book, is that there is a general misunderstanding of what types of equipment a particular facility has at its disposal. More importantly, I want to discuss the capabilities and limitations of that equipment as it relates to our ability to assist pilots.

The ATC system uses a wide variety of equipment in its day-to-day operations, and each piece of that equipment is designed for a specific use as it relates to the other equipment available at a particular facility. This equipment includes everything from radars, radios, computers, word and data processing systems, telephone systems, and plain old-fashioned people power. There are often several different types of the same basic system in use within one facility, and this equipment, including the people element, ranges in age from the newest, state of the art, elements to some that has been around for a few more years than we like to admit.

As with any equipment that is designed for a specific use, each device or system has limitations. Some of those limitations are either inherent to that equipment or are built into the device to make it useful for our needs. Let's take a look at several different systems and discuss how the pilots, controllers, and equipment interact with each other.

RADAR

Obviously, one of the more important pieces of equipment that we use in the ATC system is our *radar*.

What exactly is radar? Well, the word itself is an acronym for RAdio Detecting And Ranging. It is defined as being any of several types of systems or devices that use transmitted and reflected radio waves for detecting an object. This reflecting object could be either an aircraft or a weather system—the transmit/reflect/detect principle is the same. So radar is essentially a device that sends out a signal (a beam of energy) from a signal generator (usually attached to the receiving antenna in motion), which strikes an object, bounces back, and is picked up by the receiving antenna. (Let me remind you that radar systems are a lot more technical than the way I will describe them, but, if you are like me, the more simply they are described, the more likely they are to make sense.) The returning impulse is then displayed, using either a digital symbol or a bright spot (blip) at the appropriate location, on some type of video tube. The time it takes for the signal to make the round trip, when measured against a known element (the speed at which the signal travels), gives us the distance or *range* of the target from the antenna. Radar energy travels at the speed of light and it takes 12.34 microseconds for that energy to travel one mile and return. Technicians call this a *radar mile*. The direction from which the returning signal strikes the antenna, when measured against a known reference point (usually north), gives us the angle or *azimuth* of the target from the reference point.

What I have just described might be referred to as "passive" detection. You could draw a parallel to this concept by thinking of your eyes as a radar receiver and a flashlight as a transmitter. If you shine the flashlight (the transmitter) at a mirror (the object), a beam of energy (the light) strikes the reflector and returns to you (the receiver). In this sense the object does not have to do anything overt or active in order for that signal to be returned to the receiver except be in the way of the energy stream. This causes the signal to bounce back and the procedure is therefore described as passive detection. In this scenario, the only thing that has to work for detection to occur is the radar equipment.

But the radar system used by air traffic controllers has another type of detection capability which requires an action on the part of a piece of equipment in the aircraft to complete the detection cycle. Let's refer to this as "active" detection to differentiate it from the example of the flashlight. In this instance, the radar signal that is being sent out contains an interrogation message that, in a sense, asks the aircraft, "Who are you?" This generic question is either ignored by aircraft not equipped to answer the question, or responded to by aircraft containing an operating transponder. In very simple terms a transponder is a transmit-and-receive device that *hears* this electronic question and responds by broadcasting a specific answer depending upon what four-digit code has been selected on the face of the instrument. The electronic signal sent out from the transponder will then be received by the radar system, analyzed for which of the 4096 possible answers was given, and identified based on that analysis.

This interrogate/answer/detect procedure is our form of "active" detection and the resultant signal used to be referred to as a (SECRA) secondary radar beacon target when displayed on a radar scope. I still use that term occasionally but radar technicians now refer to it as simply the *radar beacon target*, and the concept of it as a secondary system is becoming less pronounced. We talked about this transponder identification process a little in our scenario with Ashley and J.R., and we will talk a little more about it later in this chapter, so I think that by now you should have begun to understand this active detection concept.

You might think that radar is radar, but there are actually several different types and generations of radar equipment. In the air traffic controller's world, these various types of radar systems fall into three general categories: airport surveillance radar (ASR), digitized radar and precision approach radar (PAR).

Airport Surveillance Radar

The *airport surveillance radar* (ASR) system is the type of equipment that is currently used in most terminal (tower) environments, and it uses the information gathered from a single rotating transmitter-and-receiver antenna array. This system depicts the return of an actual target on a video display unit using the primary return blip that we described earlier. In most cases this return is also associated with the secondary, or radar beacon target, collocated over the primary. We will discuss these returns in more detail later, but for now let's just associate the targets on these displays as being a depiction of the actual reflections of the target. It is an oversimplification to say that these are the "real" targets, but when we talk about processed data in the next paragraph perhaps you will understand what I mean.

Digitized Radar

Digitized radar also uses rotating antenna(s) and employs a device called a digitizer which converts actual targets into electronic symbols displayed on a radar scope. Different symbols are used to represent different types of returns on these systems, and the presentations of weather, aircraft, and obstructions are depicted at the actual locations of the real targets. This is a very sophisticated piece of equipment but it is still an artificial presentation when compared against the current ASR components. If information from more than one antenna is in use in this system, a piece of equipment called a *common digitizer* correlates all of the information from all of the antennas and displays the location of targets based on the sum total of the data at its disposal. Digitized radar is the system most widely used in the ARTCC (Center) environment. This type of radar system will eventually become the standard version used throughout the FAA as the terminal

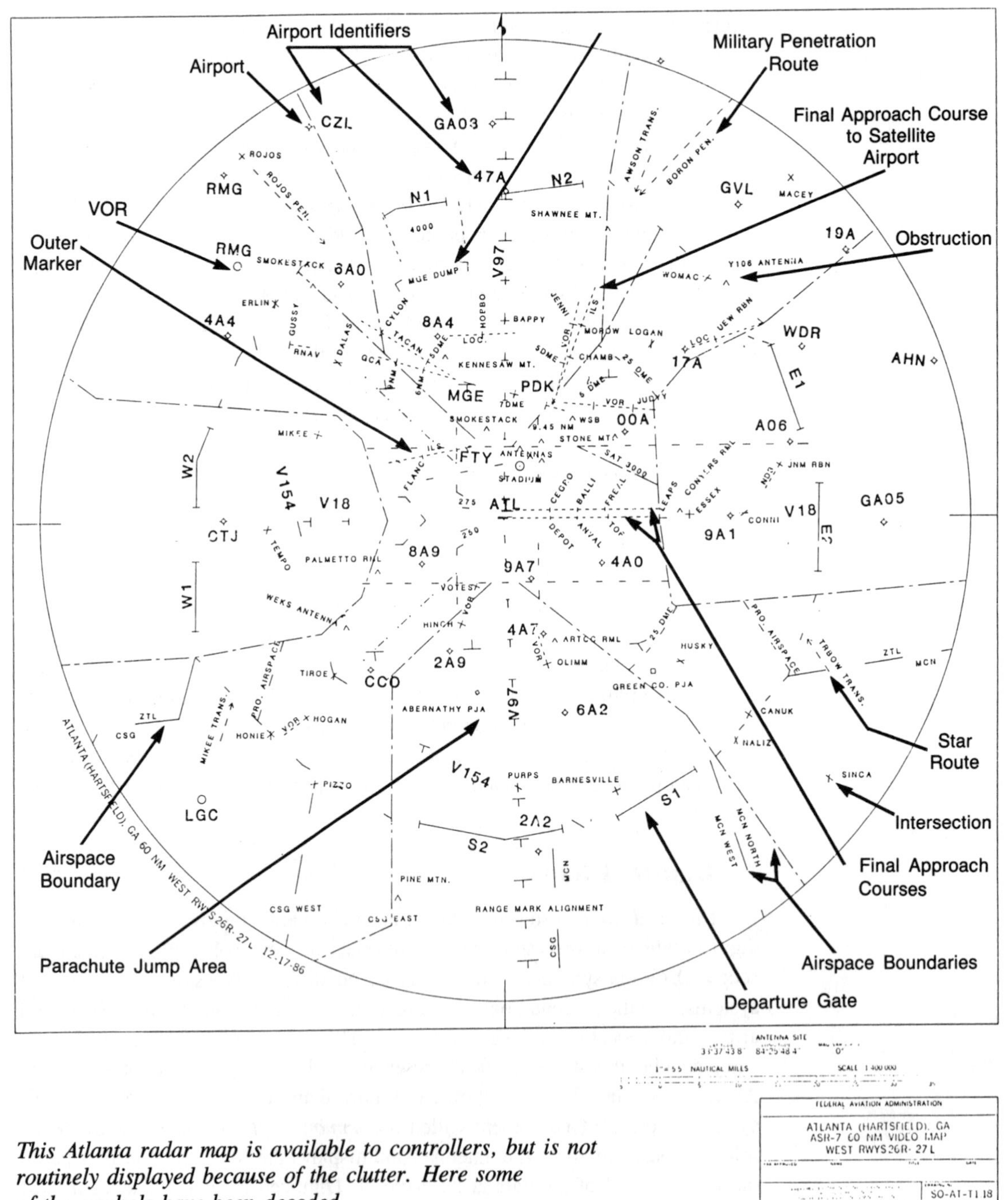

This Atlanta radar map is available to controllers, but is not routinely displayed because of the clutter. Here some of the symbols have been decoded.

facilities begin converting to the new ASR-9 radar systems which are also digital presentation displays.

Precision Approach Radar

Precision approach radar (PAR) is different from the others in two basic areas. First, it has a sweep that goes up and down rather than in a circular motion. Second, PAR systems use this vertical scan technique to determine the altitude of a target. This is done by calibrating how high the sweep will be at a given distance from the antenna and depicting a video map on the scope with a line representing a type of glide slope. This altitude function is particularly useful in the precision approach environment because the radar is aligned with the extended centerline of a runway, and the video-map glide-slope line will represent the ideal altitude of an aircraft as it descends on final. This equipment is most commonly used in the military environment as a *ground-controlled approach* (GCA) radar system, and if the controller staffing the scope at this GCA position is at his/her best, the accuracy of the approach comes close to what can be achieved by an ILS system. But the fact that precision approach equipment has a limited or fixed range of lateral (circular) motion makes it unsuitable for use in the broader range of radar services.

Of course, the primary function of ATC radar is to see and depict aircraft so that we can guide and separate them. So, any distraction from, or reduction in, our ability to do that is considered to be a limiting factor, and every technological resource available is brought to bear against these limitations. We will talk about each of these resources in more detail in a moment, but for now suffice it to say that we have several such resources and each is designed to reduce a particular type of interference and increase our ability to see the aircraft targets. They are also rather effective at doing that, so you might ask why don't we just leave all of these adjustments turned on all the time? Well, as is typical with most enhancements, there is a price to be paid when we use them. These adjustments usually reduce or eliminate some aspect of the radar return and a similar, but hopefully smaller, reduction in the strength of the radar return from the aircraft also occurs. Let's look at radar in general and several of the radar system functions and discuss both their positive and negative impacts on the controller and the pilot.

Radar Scopes

What exactly does a radar scope look like? Think of a round video display unit much like an arcade game or television screen. The scope is usually green but some are other colors. Place a point in the center representing the location of the radar antenna. Now add a reference line (sweep), emanating from that point to the outer edge of the scope, that is moving in a clockwise direction. As this sweep moves around the scope, an image (called a blip, target, or primary return) is placed on the video tube for every object that bounces back the radar signal. In this instance we are talking about surveillance radar, but the concept is the same for digital radar. The only thing different is that the blips would be displayed symbolically on digital radar.

The position where each target is displayed is relative to where the sweep is in its arc when the signal comes back, and the target's distance from the center of the scope is determined by the aircraft's distance from the radar site, relative to the range selected. For example, if we define the top center of the scope to be north and select a 60-mile range, a target that appears halfway between the center and the edge of the scope as the sweep passes top center is described as being 30 miles directly north of the radar antenna. This is a somewhat simplistic explanation, but radar is a very simple piece of equipment until you start adding all of the dealer-installed options.

Before we move on to some of those options let me again digress for just a moment to point out a critical fact related to radar returns. When the radar signal strikes a large object, such as a building, all of the radar energy emitted within the area occupied by that object is either reflected back, absorbed, or deflected away from the radar antenna. The direct result is that the obstruction shows up loud and clear on the radar and the area behind the obstruction is shadowed from the antenna. Stationary or moving objects in the shadow area will not be seen by the radar antenna.

Most people would not think that one building would create that much of a problem, but the actual problem is aggravated by the fact that the size of the shadow is determined by the size of the obstruction and its distance from the radar antenna. If you were to place a 300-foot-wide hangar two miles from the antenna, it would only shadow a very small area. But place this same building relatively close to the location of the antenna, and you might block out 3 to 5 degrees of the arc of the radar sweep and several degrees of vertical coverage. This doesn't seem like much because a degree of arc is only a few feet wide at the antenna site. But one degree of arc is one nautical mile wide when you are 60 nautical miles from the radar antenna. This 3- to 5-mile dead area can, and frequently does, wipe out the radar coverage for an entire final approach course into a small field. Depending on how much vertical coverage is also lost, you might not be able to see an aircraft in this location until it is several thousand feet in the air.

Consequently, the further you are from the antenna and the lower you are flying, the more likely you are to be within the area of one of these shadows. This problem could be partially eliminated by using more than one radar system and interfacing the multiple signals to give better overall coverage. The ARTCC radar systems normally use this type of multiplexing, but the area covered by ARTCCs is infinitely larger than that covered in the terminal environment, and the separation requirements are significantly higher when this type of equipment is used.

In terms of cost, effectiveness, and the needs of the specific operation, terminal radar environments usually use just one radar site. Unfortunately, this one terminal radar antenna is normally located on an airport and the area around most airports is cluttered with hotels, motels, terminal buildings, and hangars. Perhaps now you can begin to understand why a controller might be able to see your target at 2500 AGL in one location 40 miles from the antenna and then have difficulty finding your aircraft only 30 miles from the antenna in a different location.

Enough said on that topic, now let's get back to the first of our dealer-installed options.

Moving Target Indicator

Because of the fact that every tree, building, hill, and antenna bounces the radar signal back to the receiver and the combined display of these targets (called ground clutter) virtually wipes out large portions of the radar presentation, we use a function called *moving target indicator* (MTI) to eliminate this clutter. (This same function in the newer systems is referred to as *moving target detection* (MTD). It is somewhat more refined and sophisticated, but the concept is still the same.) In very simple terms, the MTI function eliminates ground clutter by "telling" the radar display that any object that is not moving does not exist, and therefore do not display that target. By not moving I mean that there is no relative motion, either toward, away from, or oblique, relative to the radar antenna. The speed limitation set for this motion is approximately 7 knots.

The net result is that all of the targets that meet the MTI criteria are simply cancelled and cease to be displayed within a preselected area chosen by the controller. The area chosen by the controller is variable depending on the radar scope, because at some point the function is no longer needed. As the radar beam moves away from the antenna, it rises at a preset rate and the elevation of the beam is above the surrounding terrain and obstructions. Again this is an oversimplification because the radar energy does follow a very slightly curved path due to gravity, and the curvature of the Earth causes the ground to drop away from the beam at an even faster rate. But the point where the beam is above the obstructions is usually where the controllers set the limitations of the MTI through the use of an adjustment on their radar scopes. This adjustment, or *range gate* as it is referred to in ATC language, allows them to vary the perimeter of the MTI function. Most facilities have directives which stipulate how far this function can be used, but, regardless of these procedures, the average controller tries to keep the MTI range as small as possible because we can also lose aircraft targets in this area as a result of a rather unique phenomenon called *tangency*.

If you were to draw a circle at some arbitrary distance from the radar antenna and then draw a line that was tangent to that circle, you would have just described a flight path of an aircraft that will disappear for a few seconds from the radar display because of the MTI function. When an aircraft operates on that portion of the line which intersects the circle, that aircraft has, for several seconds, zero velocity with respect to the antenna. The radar system sees what appears to be a stationary target and, according to its programming, does not display that target. Therefore every passively displayed target that flies at a right angle to the arc of the radar antenna will disappear from that radar display for several seconds at least.

It is also theoretically possible to fly a perfect circle around a radar antenna at a constant speed and never be displayed on the radar scope. You might think that it is highly unlikely that very many aircraft would fall into this category, but consider the following. Most radar systems are located at relatively busy airports, and most pilots who want to transit this airspace like to avoid the area immediately around the airport. The resultant flight path usually describes an arc around the airport/radar antenna. This arc is usually just outside the busiest area around the airport or right at the edge of the TCA. When this flight path is aided by distance measuring equipment (DME), it tends to remain almost exactly the same distance from the radar system. This kind of flight path keeps the aircraft at that tangent point as long as the pilot holds a consistent arc. This situation is aggravated by the fact that smaller aircraft, particularly cloth-covered aircraft, display a weaker target than large aircraft. Perhaps, now you can understand why some of these targets fall through the cracks of radar surveillance. I could also add to the list of those missed aircraft by factoring in that group of pilots who turn off their transponder so that they can cheat just a little. Since we all know that pilots are too smart to do such a thing, I won't mention that possibility.

While this type of target loss is the most important negative aspect of MTI, it is not the only one. MTI has other limitations which are related to weather and weather depiction and a discussion of this topic will lead us directly into a dialogue on the weather circuits used with the radar systems.

Raindrops or, more precisely, water droplets are also targets which reflect radar energy. The reflective quality of water droplets is affected to some degree by several different circumstances. Variables such as the radar signal wavelength, relative strength of that signal, and the amount of water in the air all affect how much of this weather will be depicted on a radar display. When water droplets are present in light or moderate levels and when their speed, horizontally relative to the radar antenna, is less than seven knots, they are usually eliminated from the radar presentation by MTI. The controller has the ability to turn off the MTI function so that more of this precipitation is observable, but the weather presentation may be obliterated by, or be indistinguishable from, the large area of ground clutter which also shows up during the time that the MTI function is turned off. This is why a controller may not see and report some areas of light-to-moderate precipitation within 10 to 20 miles of the antenna.

Beyond the fact that MTI eliminates some weather returns, the actual design of ATC radar does not easily lend itself to the well-defined presentation of weather systems. In very simplistic terms, the shorter the wavelength and the tighter the beam width of the radar signal, the better the presentation of small reflective particles. The signal emitted by most ATC radar systems has a rather long wavelength and a relatively wide (2 degree) beam width. Compare this with the 0.85 degree beam width of Doppler radar (which is alleged to be the greatest thing since sliced bread when it comes to weather radar), and you can see that our equipment is a rather poor detector of light precipitation. Remember, our equipment is designed to pick out the larger, occupied particles from among all of the other particles floating around in the air. Even so, as the intensity of this precipitation builds and/or the relative ground speed of the weather system overrides the MTI circuitry, the weather areas tend to "white out" large portions of the radar scope and make the aircraft targets indistinguishable from the weather return. This is what I meant when I said that tangency was not the only way we would lose targets in conjunction with the MTI function. This situation is unacceptable when you consider that we have to see the targets in order to separate and guide them so it is at this point that we start to apply other circuits designed to eliminate the weather while keeping the aircraft returns as strong as possible.

Circular Polarization

Normally, the radar antenna emits energy in what is described as a straight-line plane. More precisely, the energy is polarized, or aligned with respect to one plane or direction, and that direction is linear. This is the normal status of the radar system when there is no need to reduce precipitation returns. It is referred to as *linear polarization* (LP). When the weather returns begin to overwhelm the radar display, the controller changes to a circuitry called *circular polarization* (CP) to obtain better results. When CP is activated, the radar beam is emitted through several (usually three) tubular mechanisms which cause the beam to be emitted in a counterclockwise, screw-like motion. This action results in a significantly reduced sensitivity to spherical targets.

Water droplets, and the clouds and weather systems they describe, have a generally spherical symmetry and, because of CP, fewer of them are displayed on the controller's radar display. This function also reduces the strength of the reply from this type of target which further eliminates clutter caused by weather. Unfortunately, there are aircraft whose design is also spherical (Cessna 310 with tip tanks, among others), and the target strength of these replies is also reduced. The radar's sensitivity to all returns is reduced but, in general terms, the metal aircraft's target will still remain when large areas of the weather returns are reduced or eliminated.

LOG FTC

The controller has one more tool at his/her disposal when the intensity of the precipitation returns begin to overwhelm even the CP function. We can activate two other circuits which are called *background video circuits* and *LOG FTC*. LOG FTC stands for *logarithmic fast time constant* and an exact discussion of how these systems align radar returns to eliminate weather areas is way above my head. In terms of what a controller sees on the radar scope when these circuits are activated however, the results are dramatic. This circuitry electronically outlines the strongest edges of the areas of precipitation and reduces or eliminates returns from the areas inside those outlines. Here again, the radar simply does not depict the weaker of these returns and only the strongest passively reflected target, usually the reply from an aircraft, will be displayed on the controller's scope inside the area.

Attenuation

ATC radar has one more difficulty with the depiction of weather areas that is perhaps the most potentially dangerous of all of its limitations. Referred to as *attenuation*, it is a situation where one weather area effectively hides behind another. When a very intense area of precipitation is depicted on our radar display, it is entirely possible that most or all of the radar energy emitted in that direction is being reflected or absorbed by this system. Given this possibility, if a second weather cell exists outside of the first and in a direct line between the radar antenna and the one that is reflecting, all of the radar energy, little if any of this second cell would be displayed. To give you an example of how this situation could develop, let's consider the following possibility.

Some of the most devastating weather phenomena that a pilot can encounter are squall lines. These sometimes massive systems can block an area hundreds of miles wide across a pilot's flight path. Rather than fly the extra distance, many pilots will look for a soft spot or gap between two thunderstorms and attempt to cross the squall line at that point. If we were to build a hypothetical squall line we could begin by suggesting that a solid line of thunderstorms existed along an arc that was 35 miles west and northwest of the radar antenna (SEE FIGURE). Let's further define this system by saying that the thunderstorm line was depicted between the 270-degree bearing and the 330-degree bearing on our scope. This type of weather system is usually very prominently displayed on our radar and looks like a straight-line series of light-colored blobs on a green background.

Now let's place a small slow-moving but very intense area of weather at a point 10 miles from the antenna on a 300-degree bearing. This small area is being reduced by a combination of the MTI and CP functions and may not appear as a particularly intense weather area on the controller's radar display. If this cell

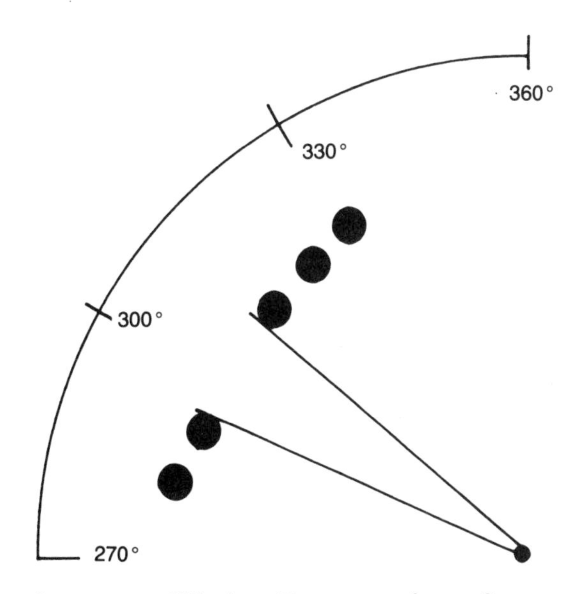

A nearby thunderstorm can "block out" coverage of more distant areas and weather systems. This phenomenon is known as attenuation.

is strong enough though, it could easily reflect or absorb the majority of the radar energy directed in its path.

The resultant display of the weather on the controller's radar scope will show what appears to be a gap in that solid line of weather located at approximately the 300-degree bearing and 35 miles from the antenna. This position is in a direct line from the radar antenna through our small, but not very well displayed weather cell. In reality this may actually be where the most intense area of weather associated with that entire system is located and this sucker hole could easily claim anyone foolish enough to fly through that point. If you ever get the chance, read the National Transportation Safety Board's accident investigation report (AAR-78-03) on Southern Airways Flight 242 which crashed in northwest Georgia after encountering extremely heavy hail in a weather area much like I have just described.

If only for this reason, controllers are very reluctant to accept the responsibility for vectoring aircraft through areas of precipitation or places which appear to be gaps in those areas of precipitation. When pilots ask us to do this, we often go into a long, drawn-out explanation qualifying our information or explaining why we cannot comply with their request. Pilots need to learn, and keep uppermost in their minds, the fact that the controller may not depict all of the weather areas that exist in their environment. We will certainly tell you about what we see, but, as you can tell from the above example, that is not always all of the information you need to make an informed decision.

Other Features

Up to this point we have only been discussing primary radar returns (actual reflections from a target), but a controller sees a lot more than raw radar returns on his/her radar display. The total presentation is made up of a combination of real-time information, video map overlays, and processed information that includes some or all of the following items:

- Passive radar reflections or what we have been referring to as primary radar returns. This type of target is only found on surveillance radar systems and is not associated with digital radar equipment. With digital radar, all of the processed information on the radar display would be displayed in a symbolic and graphic format.
- An interchangeable video map overlay which usually depicts obstructions, airports, airspace boundaries, airways and/or navigational aids, and instrument approach courses. A controller has the ability to select from the several maps, including an emergency map which shows highway markings and an exploded view of airports and emergency landing areas.
- Secondary, or active radar returns, which are electronic signals from a transponder and are referred to as *beacon replies*. The target associated with this signal overlies the primary target return from those aircraft that are equipped with and using a transponder. The resultant target is called a *beacon control slash* and looks like a longer, slightly fat-in-the-middle, primary return. (This is a good place to mention a very important aspect of active radar returns. In order for this type of target to be displayed, both radar and transponder have to be working together. If the transponder is not broadcasting an identifiable reply, it might as well be turned off. This is one reason that we sometimes cannot find an aircraft that is receiving an interrogation signal on its transponder, and it is the most important reason to keep that thing calibrated.) When this beacon control slash is displayed on a radar scope using selected beacon code enhancement features, the target has two slashes. The second of these

targets is always in a direct line between the radar antenna and the aircraft and positioned beyond the real target. The space between these slashes will fill in when the controller asks a pilot to ident. This type of system is normally only used when a facility is not equipped with or is not using a numeric or alphanumeric tracking system.

- Numeric or alphanumeric characters which provide information about the aircraft's call sign, speed, and altitude.

There are several other enhancements associated with this system so let's talk a little more about this system.

Most ATC facilities are now equipped with some type of numeric or alphanumeric tracking system that is associated with the transponder or beacon reply, and some even have systems that will track primary targets. In the terminal environment, there is a wide range of equipment sophistication in use depending on the complexity level and needs of the facility and the type of equipment that was in use at the time it was installed. This equipment is constantly being upgraded as technology improves, as FAA funds permit, and as the needs of the facility dictate.

TPX-42

One of the earlier, older systems, called *TPX-42*, uses a numeric readout on the radar scope and will only depict transponder codes and, in those cases where the transponder has Mode C capability, the altitude of a target. Another, slightly more sophisticated system, called *direct altitude and identity readout* (DAIR), is essentially an enhanced TPX-42 system capable of tracking a target and using alphanumeric characters to depict call sign, altitude, ground speed, and some additional information associated with that flight. This system is used primarily within the military and will interface with some of the systems in use at ARTCCs.

ARTS

The most commonly used types of alphanumeric tracking systems in use in the terminal environment are called *ARTS*. There are several different generations of ARTS, suffixed 1, 2, 3, and 3A, depending on the generation being used. All of these systems have a wide range of services available to the controller and, indirectly, the pilot. As an example, ARTS-3A has the capability of tracking a target while displaying, through the use of a data block which follows the target, the following information:

- the target's call sign
- type of aircraft

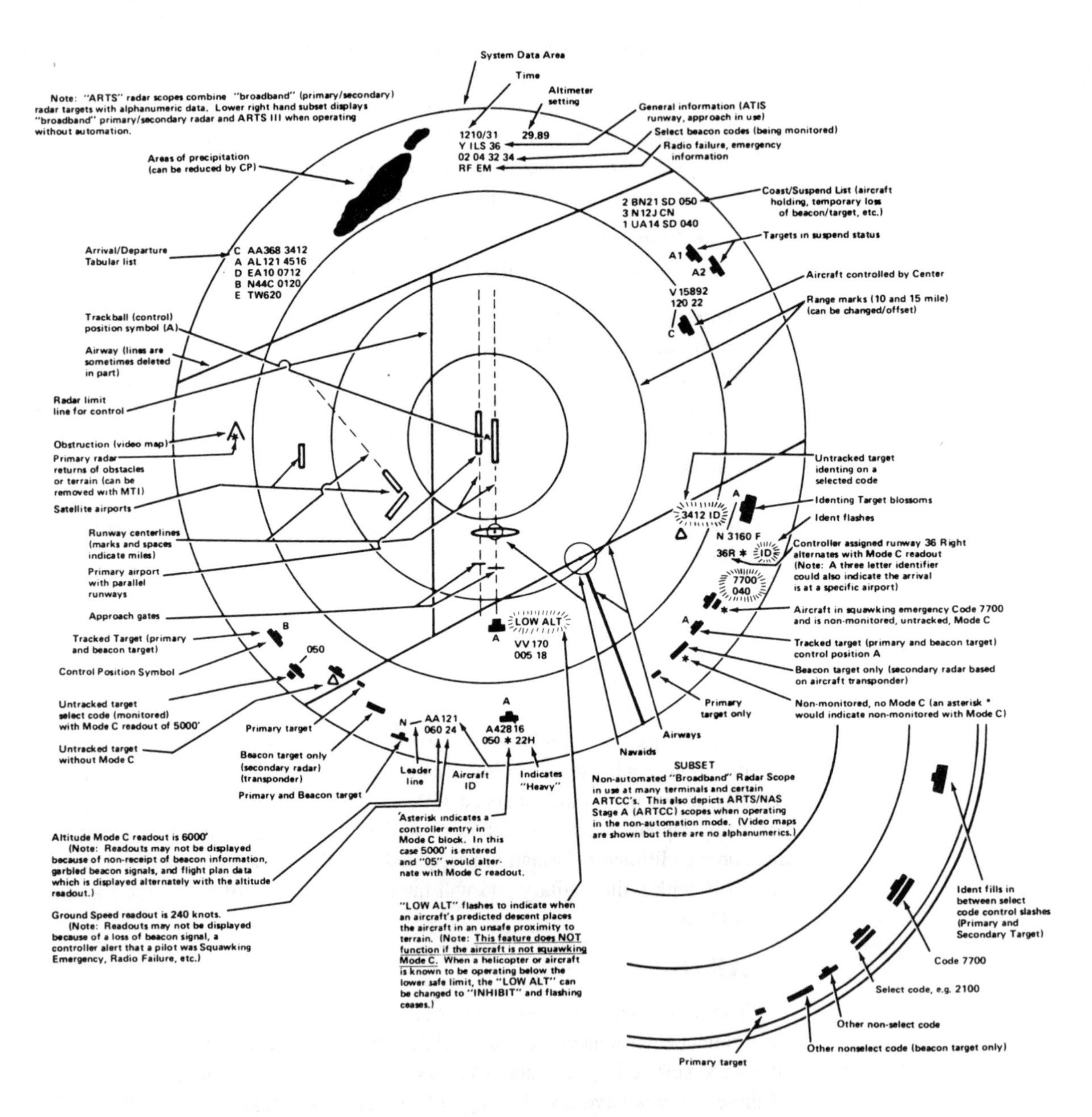

ARTS III Radar Scope with Alphanumeric Data. Note: A number of radar terminals do not have ARTS equipment. Those facilities and certain ARTCC's outside the contiguous US would have radar displays similar to the lower right hand subset. ARTS facilities and NAS Stage A ARTCC's, when operating in the non-automation mode would also have similar displays and certain services based on automation may not be available.

The ARTS III radar display.

- altitude (if Mode C transponder equipped)
- ground speed
- data concerning the type of flight (VFR, IFR, enroute, heavy, etc.)
- other information, called *scratch pad data*, which can be manually inserted into the data block by the controller.

ARTS-3A also has the capability of tracking a primary (non-transponder) target although the data block associated with this target is limited to information that is entered by the controller.

Most ARTS programs have the capability of allowing controllers to use automated handoff procedures between sectors within a facility. This is the process by which controllers transfer information concerning an aircraft or flight plan without direct voice communication. If the information concerning this flight is also entered into the host computer by the ARTCC sector controller, by the FDEP or FDIO systems, or by the FSS specialist, automated handoff procedures may also be used between other similarly equipped terminal facilities or between a terminal facility and an ARTCC.

These ARTS programs also have a variety of nationally or locally adapted features, called *patches*, which provide additional automated services for the controller. National patches are usually program enhancements which serve the same need in each facility, are generic in nature and can be installed in the facility ARTS program by simply adding local site data to the master program. Let me qualify that last statement by mentioning the fact that "simply" adding local data usually requires that a very smart computer specialist spend hundreds of hours getting this information into the program correctly.

Probably the most sophisticated and useful of these programs is the one we controllers refer to as the *emergency patch*. This patch allows us to make a few keystroke entries into the ARTS and call up all of the emergency information about an airport. At some facilities this information includes the page in the facility emergency handbooks where this information is kept in great detail. It also gives us point-to-point bearing and distance information between an aircraft and an airport or, for that matter, between an aircraft and any point in space. We can also tell the computer what category aircraft we are dealing with, and it will give us all of that emergency information plus point-to-point bearing and distance data between that aircraft and the nearest airport with the runway capacity capable of handling that aircraft. There are several other uses that this program has, but I think that you have the general picture.

This is just one of many such national programs written for the ARTS system and the only limitations are the imagination of the people writing them and the storage capacity of the computers in which they are kept.

An example of a locally adapted patch would be a program where a controller would be able to enter one or two key strokes into the ARTS and be able to generate an entire flight plan complete with call sign, type aircraft, and destination. This, of course, would only be useful at a facility which had a large number of similar operations where the general call sign, type aircraft, and destination were always the same.

During the time that I was assigned to Standiford Field (SDF) in Louisville, Kentucky, this type of program was used to identify its military aircraft operations. All of these aircraft used the call sign SKATE, and individual aircraft were identified by a two-digit identifier used in conjunction with that word. These aircraft would depart as a flight of three or four, all operating under a single call sign. When they returned for landing, they would want to break up into individual elements to conduct an approach. Considering the speed of the aircraft and the workload involved in processing four flight plans, it would have been impossible for one controller to perform this request while still controlling his/her own existing traffic.

The large number of these operations at SDF meant that a local patch needed to be developed which would allow a controller to quickly enter a local IFR flight plan into the computer. The program that was designed allowed a controller to type an asterisk plus the two-digit aircraft identification into the ARTS, producing an individual flight plan for each aircraft. For example, when a controller typed *21 into the ARTS, the computer would generate an IFR flight plan complete with a discrete transponder code for SKATE21, showing an RF-4 military reconnaissance jet as the type aircraft with SDF as the destination. The controller would simply instruct the pilot to squawk the computer-assigned transponder code, and an individual data block would be generated for each aircraft. These kinds of programs, or variations of them, are in use throughout the country.

These are just two of the hundreds of specially designed enhancements used by various facilities. These patches are typical in that they are designed by computer experts who are, or have been, controllers at some point in their careers. Many were controllers at the facilities where they currently work their computer wizardry. This gives them the unique ability to relate to the controllers' environment, see problems from their point of view, and then design a program to solve that problem. As a result, those terminal FAA facilities equipped with ARTS systems usually have a combination of national and facility patches designed to reduce controller workload associated with some type of frequently used local aircraft operations.

As a terminal controller, I am most familiar with the type of systems in use in my portion of the ATC environment, but the ARTCC facilities also have alphanumeric systems and computer patches which are based on the same criteria. Similarly, these programs were created by computer and equipment experts who usually have ARTCC controller experience in their background. In general terms, the equipment in use at ARTCC and terminal facilities performs the same functions, but uses different kinds of brand-name equipment to get the job done.

Pilot Awareness

Pilot awareness of this equipment and the patch enhancements can often reinforce their training, reduce coordination time between pilot and controller, and in some cases even save their lives. Our facility recently recorded a save where a pilot lost his engine, transponder, and transmitter in IFR conditions over some rough terrain. The controller, using the emergency patch described earlier, began issuing bearing and distance information to the nearest airport to this pilot by transmitting in the blind. By keeping up a constant stream of information, the controller and the pilot were able to guide the aircraft to a modified final approach course which allowed the pilot to see the airport just as he broke out of the low overcast. The pilot later stated that he would not have made it safely onto the ground without this guidance. He believed that he would have crashed into a heavily wooded area because he had only seconds to choose a landing site after he broke out of the overcast. Had it not been for the help and the accuracy of the information provided by the controller and the pilots ability to react to that information, I believe some lives would have been lost.

Most pilots are totally unaware of what capability exists within the facility that controls the airspace around their home base and, because of this ignorance, they fail to take full advantage of all the services at their disposal. This lack of knowledge and the resulting inability to take advantage of that information can be very costly. In the above example it meant the difference between having a war story to tell your grandchildren and becoming a footnote in some aviation accident publication.

RADIO AND TELEPHONE SYSTEMS

Up to this point we have been talking about the radar and all of its enhancements, attachments, and associated equipment. While this is one of the more important tools at the disposal of the air traffic controller, it is by no means the only one used.

In a radar environment, probably the next most important equipment is the communication system that controllers use to speak with each other and to the pilots. The telephone and radio networks, which are the two devices that make up our communication system, are completely separate systems. But they are so interconnected and so often used interchangeably that the controllers tend to think of them as one communication package. I would like to discuss each of them individually and in depth, and then show some examples of how pilots and controllers can get very creative and use one in place of or to enhance the other.

Radio Systems

The radio systems that a controller uses to communicate with pilots vary according to the size and age of the facility. Some are very simple two- or three-channel transmit/receive comm sets with one-hand held microphone jack. Others are very complex multijack, multichannel, digitized star wars systems.

The standard design of these radio systems is one that establishes a discrete frequency for each position of operation within the facility. The transmitter/receiver *key packs* located at each position usually have space in them for six to eight plug-in modules which, when connected to a master frequency-selector control panel, allow a controller to broadcast and receive on a particular frequency. The primary frequency assigned to a given position normally occupies the number one (farthest left) slot in that key pack. These frequency modules are interchangeable, usually equipped with a push-button or microswitch selector to turn them on and off, and they are generally connected to another switch which allows a controller to use either the primary or backup system of transmitters and receivers. (Almost every radio communication package in use in the FAA has a completely redundant system connected to it with another full set of transmitters, receivers, and antennas. Additionally, most FAA facilities also have a portable battery-powered backup system for all or most of the frequencies in use in the facility.) The modules, or whatever other type of transmit/receive system is in use, have some type of an indicator which tells the controller that the frequency is in use. The modular key pack in use at Atlanta, for instance, has a light that flickers when someone, pilot or controller, is transmitting on that frequency. This could be very important information when a controller is broadcasting on more than one frequency. (More on multiple frequency usage in just a moment.)

While we are talking about radio frequency assignments, let me make one point about the use of radio frequencies that a lot of pilots obviously do not understand. There is a finite limit to the number of VHF or, for that matter, UHF radio frequencies that can be used in the ATC system. The fact that there are more control facilities and more individual sectors than there are frequencies creates a serious problem.

Let's begin with the premise that the 360-channel radio is the type of equipment that most aircraft will have as minimum standard. While there are other, more sophisticated VHF radios capable of channel selection in movements of .05 MHz, not all aircraft are equipped with this equipment. Those that are will normally be found at the busier airports or in the higher elevations of the ARTCC airspace. Given this situation, each ATC facility must retain the capability of transmitting in the 360-channel range on some, if not all, of their sectors.

Frequency Management. Radio transmission signals are omnidirectional line-of-sight radio waves. As such, the effective range of these transmissions increases as the altitude of the aircraft increases. For this reason we are unable to assign the same frequency to two facilities that are within a few hundred miles of each other when they control aircraft at relatively high altitudes (10,000 feet and above). But we can, for example, give the same frequency to two VFR towers facilities that are less distance apart if these facilities are only controlling aircraft in the pattern, and their aircraft are well out of line-of-sight range of each other. The assignment of this frequency to both of these airports is effective frequency management, and it will release another frequency back into the pool for assignment elsewhere.

The procedure by which each facility is assigned a particular frequency is a rather complex system of frequency management. The procedure is complicated by the fact that a large number of the available VHF communication frequencies are set aside for specific use other than the direct control of air traffic. Some of the frequencies that are set aside for specific use are the VHF emergency frequency of 121.5, the several UNICOM frequencies and the flight service discrete frequencies. The actual frequency assignments allocated to a particular facility are designed to ensure that no pilot receives a transmission from a facility other than the one to whom he or she should be speaking.

Unfortunately, Top Gun and his sidekick J.R. are always out there, and their need to talk to the controllers at the small airport is more important than anything else that may be going on around them. (Actually, they are probably too dense to realize that another facility might also be using this same frequency.) So while they are cruising along at flight level 290, 500 miles from the little airport, they select the tower frequency, or worse yet, the ground control frequency and give us a call. I mention the ground control frequency as being worse because these frequencies are assigned at much closer intervals. It is expected that the line-of-sight transmissions on this frequency should be limited to the range obtainable by the height of the control tower.

Top Gun and J.R. begin a long-winded dissertation of who they are, where they are (they are very proud of the fact that they are up there with the big folks), and what they want. While this is going on, the controller at the very large airport, who is working all of the traffic that the law allows, is quickly going down the tubes because the frequency is blocked. By the time that the controller can get this clown off the frequency and regain control, the airshow would be worth the price of admission. Getting these guys off the frequency usually takes two or three transmissions, at least one of which has to be a threat.

The point to be made for all of you aspiring Top Guns is that specific frequencies are set aside for you to obtain the information that you desire. If those frequencies are busy, wait your turn and do it right. Remember, a lack of planning on your part does not constitute an emergency on ours. If you persist in tying up our frequency, we will be happy to have a representative of the FAA meet you at your destination airport and fully explain the ramifications of your actions.

Those individuals might even be willing to provide you with 30 days of leisure time so that you can learn and contemplate the correct procedures.

Key Packs. Getting back to the transmit/receive key packs, there are still five to seven modules available at each position after the primary frequency has been assigned. For the sake of simplicity, we could call the position that we have been discussing Position No. 1 or P1. Normally, the remaining modules in the key pack at P1 would contain the same frequencies allocated to control positions operating nearby. For example, if P1 were a local control position, one of the modules in P1's key pack would contain the frequency assigned to the ground control position (P2) associated with that local control position.

The logic under which we are operating is that as the traffic decreases, control positions are combined and the position taking control has to have the ability to talk to aircraft who would normally call on the position that is no longer staffed. Can't you just visualize a controller running around a radar room grabbing a microphone at each position, talking to one pilot, then running to the next position while trying to maintain some idea of what is going on. That might be good for working off the spare tire that some of us carry, but it doesn't make for good ATC.

In addition to the primary VHF frequency assigned to each position and some of the frequencies of nearby positions, each key pack usually contains a UHF frequency. Depending on the type of traffic controlled by that position, that frequency may be discrete to that position or may be one that is shared by several of the same type of positions. For example, in the Atlanta TRACON all of the positions that normally work military traffic in and out of Dobbins AFB have a discrete UHF frequency assignment. All of the positions in the tower cab, which rarely work UHF traffic, share one frequency which is selected by the appropriate controller when necessary. Some key packs, usually those positions to which several other positions are combined, also contain the VHF and/or UHF emergency frequencies so that a controller has ready access to any pilot who needs help quickly.

Most small FAA ATC facilities are able to fit all of the frequency requirements for each position into the typical radio key pack. Unfortunately, some of the larger ATC facilities also have other kinds of frequency management requirements that have to be considered in the assignment of which frequency goes with which module. This occasionally forces a choice among several different sets of priorities.

It is important to note here that we are not talking about equipment limitations or the failure of the FAA to provide the controller with the needed tools to do his/her job. It is entirely possible to build a system that allows one controller the capability of talking on every possible position's frequency. But, if we were to do that in a place like Atlanta, that controller would then have to be able to respond to information coming to him/her from as many as 25 different primary VHF and UHF frequencies. The limitation in this situation would most likely come from information overload if the controller tried to sift through what was being said by 25 people calling at the same time.

Let's talk about some of the conditions which bring about frequency assignments to a particular position(s) and then discuss the physical layout of typical facilities in order to explain why some controllers end up working several frequencies. Hopefully, this discussion will help explain the impact that multifrequency capability on the part of the controller has on the pilot and that controller. This latter aspect of frequency usage is where old U. R. Smart got into trouble.

System Improvements. One of the biggest improvements in the communication systems that we currently use is our ability to talk to a pilot who is sitting on the ground at a remote airport several miles away from where the controller is located. Generally, there are two ways to accomplish this.

First, a remote transmitter/receiver site is located on top of a building or hill that is in direct line of site between the primary facility and the remote airport. The pilot transmits to the remote unit and the information is carried to the control facility via telephone land lines or microwave antenna. This extends our range of coverage and allows the pilot to obtain a clearance or cancel IFR without the time delay associated with going through FSS.

The second arrangement involves actually taking control of a frequency that is assigned to another ATC facility when that facility closes down operations for some reason. In the Atlanta area, there is a small but, during daylight hours, very busy airport called Dekalb Peachtree Airport (PDK). This facility does not operate 24 hours a day because the traffic operating into and out of the airport during the midnight shift does not justify the cost of operating the tower. To accommodate the IFR traffic that does wish to operate during this period, Atlanta approach control has the ability to select the tower frequency and talk to aircraft on the ground at PDK. We do this by means of a remote switch which activates the transmitter/receiver equipment located at PDK. While we do not actually control traffic on the airport using this frequency, we have the ability to issue IFR clearances, hear pilots cancel their IFR flight plans, and generally communicate the types of information that make the system more effective. The ability to conduct this type of operation also requires that some position(s) in the ATL TRACON have this frequency module available in their key pack.

Joint-Use Frequencies. Facilities that control military traffic frequently have to vector aircraft to a point where military GCA controllers take control of the aircraft and vector them for a precision approach. If the aircraft were handled in a normal fashion, this would require three frequency changes within a relatively short period of time. Considering the speed of these aircraft, the pilot would be constantly changing frequencies with the last frequency change occurring at the most critical phase of the flight. Additionally, military procedures stipulate that aircraft on a precision approach be assigned a discrete frequency so that there is no chance that communications intended for one aircraft will be mistakenly received by another.

To meet the military requirements and eliminate the problem of multiple frequency changes, a frequency, may be set aside which allow the aircraft to conduct a *single-frequency approach* (SFA)—an approach to landing without having to change frequencies. Essentially, the aircraft is assigned one frequency and different people take control of that frequency as the flight progresses. Those terminal approach control positions which would initially control these aircraft and vector them to the GCA pattern would have to have the capability of transmitting on all of these frequencies. (In the Atlanta approach environment there are three such frequencies which occupy three of the eight modules in that position's key pack.) The GCA controller also has the capability of transmitting on these same frequencies and once this controller has established radar contact with the aircraft, he or she simply takes control of the frequency and commences vectors for the precision approach.

What I have just explained to you is a general overview combined with a couple of specific examples of why certain positions might be assigned a particular set of frequencies for their use. Now let's take a look at how one or two *master positions* (positions to which several other positions are eventually combined) are created. When we're done you will know how they are set up and understand how you can make the task of establishing communication with the right controller on the right frequency a lot easier for all concerned.

Facility Layout. Although the actual physical layout of each facility's operating quarters may vary, the standard arrangement is to try to group together those positions of operation that are doing basically the same thing so that some of the equipment that is specific to that function can be shared by all positions doing that job. Let me explain this concept by describing how Atlanta Approach Control is set up and by making a comparison between it and other facilities so that you can get the general picture of who works what.

The operating positions in the Atlanta TRACON are divided into three separate functions (SEE PHOTOS). We will talk about individual position assignments in a moment, but first let's talk about general position functions.

- All of the radar displays and associated handoff positions on one side of the room control only traffic that is flying into the Atlanta Hartsfield airport. This is referred to locally as the "approach wall."
- The positions on the other side of the room are broken down into two different functions:

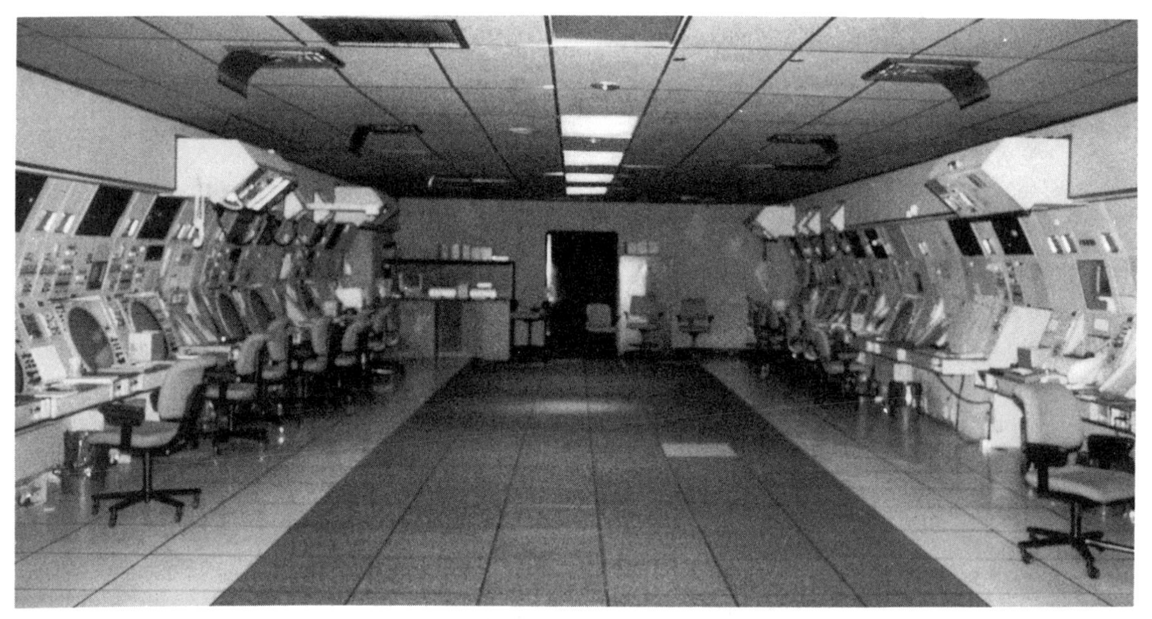

Atlanta TRACON. At left, the "approach wall." At right, the "departure/satellite wall."

The "approach wall" at Atlanta TRACON. Here all arrivals for Hartsfield airport are handled. FDEP and FDIO equipment is at far right.

The "departure/satellite wall" at Atlanta TRACON. The two positions at right handle Hartsfield departures. The four positions at left handle operations at other Atlanta area airports.

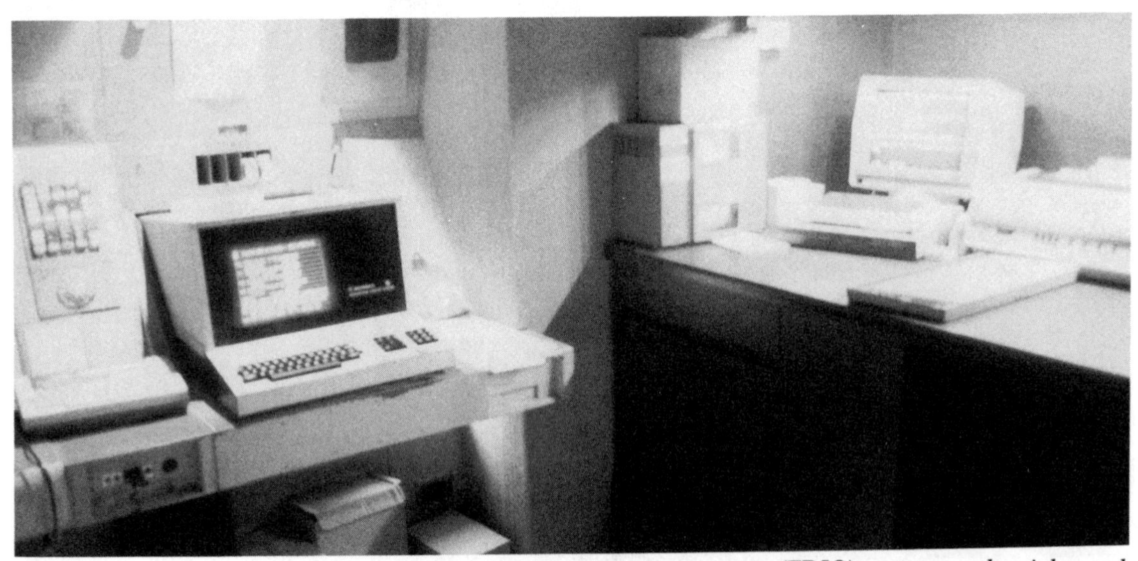

The corner of Atlanta's radar room houses the flight data input/output (FDIO) system on the right, and the Information Display System (IDS) on the left. A telco key pack is visible at far left.

- Four radar displays and associated handoff positions on this side are assigned to what we call the *satellite* function. This function controls traffic into and out of all of the airports other than Hartsfield within a 40-mile radius of Atlanta.
- Two other radar displays and associated handoff positions on this side are used to control traffic departing the Hartsfield airport. We refer to the equipment on this side as the "departure/satellite wall."

Most facilities like to subdivide their airspace or position responsibilities according to some logical plan based on what is available at that facility. They may extend the centerline of the main runway at the primary airport to the edges of the radar scopes and assign one half of that airspace to one controller and the other half to another. Or they may divide the responsibilities by assigning all arriving aircraft to one controller and all departing aircraft to another. If the facility has a VOR on the airport and VOR airways that bisect their airspace in a logical fashion, they may subdivide their airspace according to the areas described by these airways.

The plan for each facility is different, and some facilities may use more than one of these demarcation concepts depending on the positions, the workload, and/or the time of day. Atlanta, for example, uses all three of the techniques I described above to split up their operation. Our facility has a dual parallel east/west runway configuration, and we divide our airspace into north and south semicircles for all three of the functions. We also have a north/south VOR airway that further divides our airspace into quadrants. Depending on the traffic workload we can split the operation into two to four segments and assign aircraft operating in these airspace segments to different controllers. We further segregate traffic by splitting arrival and departure operations at Hartsfield into separate functions, each with its own set of frequencies.

The control tower cab is also divided into several segments. We have a ground controller who works traffic that is north of the tower building and one who works ground traffic that is south of the tower building. We have a local controller who works arriving traffic on one of the north runways and a different controller who works traffic that is departing from the other north runway. We also have two other controllers who do the same thing on the south complex. This is what we refer to as splitting the local control positions four ways. When weather and traffic conditions permit, we combine the local control positions together to form one north and one south local controller. We also have one controller who issues clearances and another position staffed by the supervisor of that function.

Sound really confusing? It's not if you look at the logic. Let's assume that you are N123TA and are southeast of Atlanta inbound to a small airport that is northwest of Atlanta. If the satellite function is very busy and is split four ways, you would talk to ATL on frequency 132.55. This is the frequency assigned to the controller who owns the southeast quadrant of satellite airspace. Now let's say that the traffic on the south side of the airport has died down to the point that we no longer need two controllers to handle this airspace. Our procedure is to combine the southeast quadrant airspace with the southwest quadrant to form a new "south" sector. This new mini-master position is operated from the control position where the southwest controller was stationed, so he or she continues to use the primary frequency assigned to that position as the primary for the new south sector. N123TA should now call Atlanta on 119.8, the old southwest sector frequency.

The south sector controller has to have the capability of broadcasting on both of the south sector frequencies so that he or she can communicate with a pilot who, unaware of the resectorization, calls on 132.55. Normally, when a pilot calls on the wrong frequency as a result of this resectorization, the controller will simply key up that transmitter and ask the pilot to change to the correct frequency. As the traffic dies off even further and we continue to combine positions, we will eventually combine down to one satellite position which, in our operation, is the northeast sector on frequency 119.3. So N123TA, calling from the same location at different times of the day, could talk to Atlanta approach on any one of three separate satellite frequencies. If we have combined further (down to one or two sectors) on the midnight shift (we will talk about this in a moment), the controller may be assigned a forth or fifth frequency.

The approach control function, which is vectoring aircraft into the Hartsfield airport, works in much the same way. Arrivals from the north initially talk to ATL on 126.9. They are then changed to the north final approach controller on frequency 127.25, and from there to the north arrival local controller in the tower cab on 119.5, and finally to the north ground controller on 121.9. Those inbound from the south follow the same basic path beginning with south arrival 127.9, south final on 118.35, south local on 119.1, and finally south ground on 121.75.

As traffic dies down, the approach control positions gradually combine together to where we use one arrival position and one final position (the two south positions). Eventually, as traffic dies down toward midnight, we combine everything to the south arrival position. This master position has to have the capability of transmitting on all of the other frequencies, because pilots are creatures of habit and often will call on what is the normal frequency during busy times of the day.

Similarly, positions in the tower cab gradually combine to the point where we have one local position at the north local and one ground position located at the north ground. If necessary, these positions can also be combined together at the north local. This master position has the capability of transmitting on virtually every frequency in the tower cab.

The third function in the ATL TRACON is the function which contains the departure positions. Traffic departing the north runway complex will normally talk to controllers on frequency 125.7 and those from the south complex on 125.0. (We also have the capability of establishing a west departure position but, in our current configuration, this scope is assigned as a training position.) When traffic is light, these two/three positions combine to the north departure position which, as you will soon see, becomes the master position in the TRACON during the midnight shift.

Weather and traffic delays permitting, shortly before midnight each evening we have combined our positions down to the point where we have one satellite position (the northeast on frequency 119.3), one arrival position (the south on frequency 127.9) and one north departure position on frequency 125.7. Depending on traffic, these positions will combine together to the north departure position creating one TRACON master position for a few hours during the light traffic on the midnight shift. This position has to have the capability of selecting most of the frequencies available to any other position in the TRACON without having to physically move to that position. Since each key pack only has eight possible modules, we solve that problem by giving the north departure position the option of selecting all of the master frequencies from the other functions. We then make most of the rest of the frequencies selectable at the south or west departure positions which are only two feet away from the north departure position. If you call on one of the incorrect frequencies, the controller simply picks up a hand mike plugged into another departure position and instructs you to call on 125.7.

What I have just explained to you is the combine/decombine sequence at Atlanta. Other facilities use similar procedures to assign airspace to individual positions, and some even combine the radar function to the tower cab during the midnight shift. We all can, and frequently do, combine and decombine positions several times during the day when traffic permits. Few, if any, facilities have a lot of spare bodies lying around, so we do this to, among other things, align the people to the traffic, facilitate coffee and lunch breaks, perform training, and meet the individual currency requirements on all positions.

The main reason that I have gone into such detail about these procedures is to show you that you need to be aware of the fact that communication with the right controller might occur on different frequencies depending on what situation exists at that time. The big question that always comes to the pilot's mind is, "How can I possibly be expected to know what frequency to use if you keep changing the game plan?" The answer is really quite simple, use the one that you would normally use. As I have shown you, the controller who now controls that airspace will usually have that frequency available or at least a means to reach that frequency without too much trouble. If a position has been combined, the controller will give you a new frequency assignment.

The next logical question is, "How do I know what is normal?" Again the answer or, if you will, answers are simple.

First, read the charts that you should have for every flight that you take. IFR, VFR Sectional, and TCA charts will tell you what the correct frequency should be if—and this is a very big if—you only use current charts. In addition to these charts, the information can be obtained from the *Airport/Facility Directory* and, in some cases, from the ATIS that is being broadcast by a facility.

Second, obtain the information from another ATC facility. If you are on the ground at an uncontrolled airport, spend a quarter or whatever it takes to contact the local FSS and obtain the information from them. If you are on the ground at a tower-controlled airport, ask the controller for radar services. This action will ensure that you will be given the appropriate frequency. If radar services are not available for whatever reason, ask the controller for information about which frequency to monitor.

This same technique applies if you are currently being given radar services. We always advise surrounding facilities when we reconfigure our operation, and they are aware of the correct frequencies for use in a given location. If they transfer radar identification of your aircraft to our facility, they will tell you what frequency to use. If they simply terminate radar services, ask them for the next appropriate frequency. The worst that can happen is that the controller will be too busy to work with you and you will be able to listen on the appropriate frequency and obtain a general picture of what is going on by just paying attention. When, and if, the traffic dies down, ask the controller if he or she now has time to provide you with radar advisories.

You now have or know how to obtain all of the information you need to acquaint you with the procedures that controllers use to establish frequency assignments. Hopefully, I have made the point emphatically enough that you should become aware of this information before you blast off into the wild blue yonder.

Remember pilot U.R. Right in an earlier chapter? By equating that situation to the Atlanta airspace you can now see the following. Because of the fact that it was after midnight when the aircraft arrived in the ATL airspace, this aircraft was probably vectored through our airspace while transmitting on frequency 125.7 and being worked by the north departure controller which was the only operating control position at that time. In fact, that aircraft might have been the only aircraft on the frequency. Let's suggest that he or she could have been going to the Fulton County (FTY) airport near the Six Flags amusement park northwest of ATL. This nice easy transition obviously lulled the pilot into believing that the outbound trip would be the same.

When the pilot left for home, he or she should have asked FTY Tower for radar advisories, departed east or westbound, and talked to the northwest satellite controller on frequency 121.0. Instead the pilot departed in a southeast direction, which immediately penetrated the TCA and the very busy north final approach controller's airspace. The pilot then compounded the problem by calling the north departure controller who knew nothing about the aircraft. Whether this situation was hypothetical or not, Smart's operation was potentially disastrous and is simply the result of a pilot who has not taken the time to learn the correct procedures.

There are very few locations in this country where you are not within the radar coverage area of some ATC facility. We are there to provide ATC services and will continue to be there whether you properly use the system or not. The frequency of occurrences such as the Smart family's odyssey indicate that there is a large contingent of pilots flying in the system who have little or no idea of how that system works. Your failure to use these services, especially if you use the excuse that you do not know how to contact the appropriate controller, is an inexcusable waste of resources. It also indicates a failure in the learning process by which you should have been made aware of this information. Hopefully, the last few pages will help eliminate part of that problem.

Telephone Systems

Now that we have covered the subject of frequency assignments like a blanket, let's talk a little bit about the controller's telephone communication capability. Other than the situations where a pilot uses telephones to obtain a clearance, the average pilot rarely becomes directly involved with the types of systems that controllers use to communicate with each other. What pilots do not realize is that the vast majority of ATC operations involving aircraft movements, including the movement of their own aircraft, require some type of telephone communication between two or more ATC specialists. Additionally, a thorough understanding of how this system works will eliminate some of the irritation that some pilots feel when the controller does not immediately respond. This understanding may also provide them with another tool in getting their job done.

Telco Key Packs. Virtually every control and handoff position in every ATC facility is equipped with a *telephone key pack*. The size and complexity of these key packs is dictated by the needs of the position. These may range from a small 8-button unit with very limited communication capability to a 50-circuit key pack that connects the controller with every position in the house and numerous other facilities in the area. There are six standard types of circuits that are commonly used in these systems and an individual key pack might have any or all of these systems contained within it. The six are referred to as radio position circuits, override circuits, dial-code circuits, push-to-ring or ring-down circuits, hot-line or voice-call circuits, and monitor circuits each has specific functions, so let's cover each one individually.

All control positions that have frequency selection and radio transmit/receive capability have a circuit that we refer to as the *radio position* on their telephone communications (telco) key pack. This circuit connects the controller to the radio transmitters and receivers through the telephone key pack. Using this system allows a controller to use the same headset for any communication associated with that key pack. (Those positions that do not have radio capability have a release key located in the same position as the radio key so that the action of terminating a telephone message is always the same.) This radio position button is the normal operating position that a controller on a control position uses when communicating with pilots over the radio, and the selection of any other key on that key pack deselects the transmitter so that the controller is not broadcasting telephone messages over the air. Any broadcasts that may occur from the pilots during the time that the controller is on another line are rerouted through a speaker directly above the control position so that the controller can hear what is being said. There is, of course, the possibility that a pilot may call while the controller is listening to something on another line and that the controller would not hear what was said by the pilot. Usually though, the controller knows that someone said something, and when they return to the radio position, they will ask for a repeat.

The *override circuitry* is actually two separate functions depending on how the equipment is being used, so allow me to explain the difference. Almost every position in use in the FAA has two equipment jacks that allow two completely separate sets of headphones to be plugged in at the same time. These positions are wired so that one of these jacks is a control or supervisory position and the other is a slaved or training position. In a training environment, the supervisory position completely disables the training position when it is activated so that the individual conducting the training can simply override the student should the situation arise where this is necessary. We conduct a considerable amount of controller on-the-job training in the FAA, and this capability assures the pilot of the best and safest possible air traffic control services while this training is being conducted.

This same type of supervisory position arrangement can also be wired into two completely separate control positions so that one has override capability over the other. Pilots will normally only encounter this type of situation at facilities which conduct simultaneous ILS approaches to parallel runways. These facilities use a monitor position at which controllers watch the flight path of aircraft on the final approach courses. These monitor positions are wired so that the controllers who staff these positions have override capability over the local control position on which the aircraft are listening. Should one of these aircraft stray off course toward an aircraft on the other final, the monitor has to have instantaneous communication capability. When they key the transmitter at their position, all other controllers on that frequency are overridden.

The second type of override capability at the telco position is a more passive type of operation and is generally an in-house type of communication. The key packs at most control positions contain a selection of keys that link that controller with other control positions nearby or with those positions with which a controller must routinely coordinate. When that controller needs to talk to another controller at a different position, he or she simply selects this override circuit and is immediately connected to the receiving controller's headset. To talk to that controller, the initiating controller simply keys their mike in a normal fashion and begins talking. When this circuit is activated, the receiving controller sees a light illuminate on their key pack and notices a difference in the side tone or background noise in their headset. Normally, this controller need take no further action to communicate with the caller other than announce their readiness to begin communicating as everything they say can be heard by the caller. (Most facilities have set procedures for how these communications should be conducted so that there is no misunderstanding as to the intent of the communication or with whom the communications are being conducted.) The following would be a sample override conversation between two controllers.

Caller: [Activates override circuit.]

Receiver: *Go ahead override, this is North Final.*

Caller: This is South Final, verify the altitude on Tandem 343 please.

Receiver: *The altitude on Tandem 343 has been verified and is currently reading 5300 feet.*

Caller: I'm a little high with Tandem 98, so I'll stay a thousand feet above Tandem 343 until I clear him for an approach.

Receiver: [States operating initials and usually says] *Approved as requested* [if agreeable to the operation or adds conditions to the request if necessary].

Caller: [States operating initials and responds to any conditions as needed.]

In this instance the caller needed the altitude verification to separate one of his/her aircraft from Tandem 343 using altitude separation. It is probable that the caller's aircraft would be less than three miles from the other controller's aircraft before one of them was cleared for an approach. Rather than delay the aircraft, the controller coordinated an alternative form of separation using the override circuits.

The *dial circuits* that terminate at an operating position are similar in many respects to the telephones that you have in your home. In fact, some facilities actually have commercial telephone lines terminated at a control position so that pilots can obtain clearances directly from the ATC specialist while calling from the pilot lounge at the airport. In most cases though, these dial circuits are more similar to the interoffice phone systems that you might see at work. When we activate one of these lines by selecting the appropriate button on the key pack we must then dial some combination of two to four digits on the rotary dial to reach the desired control position. This type of system has several advantages, and unfortunately, two major disadvantages. The large number of possible termination points allows a controller the ability to talk to almost any control position in any of several nearby facilities. The controller, of course, has to know the correct dial code for that position but that is easily solved by keeping a list of positions and dial codes nearby. Additionally, most positions only have a need to talk to a few selected sectors that normally interact with a given control area, and these are quickly memorized. But these circuits are like the old party lines and are frequently busy when we might want to use them. The other major disadvantage is that, unlike voice-call lines or override circuits, once these lines are activated and in use, a controller cannot just punch off the line to talk to a pilot and then punch back on to continue the conversation. Once the connection is made, any disconnect requires that you reestablish communication by redialing the number and verifying that you are talking to the same person that was previously on the line. As a result, these lines are most frequently used by controllers on flight data or handoff positions. If they are being used by a controller on a control position, there may be a period of several seconds where they will be unable to respond to a pilot who is calling on a frequency. Of course we all know that pilots have the patience of Job and are willing to wait for the controller to return to the frequency before becoming irritated at this inattentive controller, so this is never a problem.

The *ring-down* or *push-to-ring circuits* serve the function of connecting several different positions to one main individual or position. The receiving position in these cases is usually either a supervisory position, a flight service or National Weather Service master briefing position, or similar type position where numerous control positions would have cause to contact this individual. The individual initiating this call simply presses the appropriate button on his/her key pack and this action causes a phone to ring or buzzer to go off at the desired location. If the phone at the receiving end of the line is also on a key pack, a light will flash or illuminate indicating which circuit is being called. This type of circuit may be a one-way call line out from a control position, or it may be associated with a different type of circuit from the other direction. In this instance it could be a push-to-ring line from location A to B and a dial circuit from B to A. Like the dial circuits described earlier, the controller must stay on the line once this circuit has been activated or they will have to reestablish communication.

A *hot-line* or *voice-call circuit* is the type of communication where a controller speaks into a line and his/her voice comes out of a speaker at or near the position of a controller with whom they wish to speak. There are usually several of these types of circuits at each control position and each has a circuit number assigned to it that specifically identifies which line is being used. These circuit numbers are those used by the telephone company to identify the line and are usually six to ten alphanumeric characters long. Controllers use the last two or three digits of these numbers to associate the line number to the individual key being used. For example, the 40GP18226 line would be referred to as the 226 line. The controller, through training and experience, knows which sectors have those same lines terminated at their positions, and he or she communicates by calling the desired sector and announcing on which line the call is being made. A sample call would be as follows.

Caller: [Selects the appropriate circuit, pushes the key on the key pack and on their handset transmitter, and says] Atlanta Center north departure sector, Atlanta Tower on the 226 Line.

Receiver: [Selects the appropriate circuit, pushes the key on the key pack and on their handset tansmitter, and acknowledges the call.] *This is north departure, go ahead.*

The voice-call line is most frequently used for calls between two different FAA facilities since it is easier and less costly for several positions to share one voice-call circuit than it would be for discrete, individual lines to be established to every possible position of coordination.

A hot line is similar to a voice-call line in that the voice of the individual initiating a call goes out over a speaker at another location. They differ in that a hot-line caller must continually depress the button while talking, but does not need to depress the handset transmitter key to broadcast the information. This type of circuit is most commonly used internally in a facility, often from the tower cab to the TRACON and/or the reverse, and is used to announce generic information or broadcast messages for general consumption. In Atlanta, the flight data position uses the hot line to announce changes in the ATIS code so that all positions can hear the information and advise pilots of the change. We have a hot-line speaker centrally located on each wall.

The last type of circuitry that I want to cover and one of the more useful types of equipment in our telephone systems is the capability to monitor a control position from one or more different locations. The *monitor position* on a key pack is usually found at a handoff position or supervisory console, and it allows the capability of listening in on everything that is being said on a selected position(s) without interfering with the communication capability on that position. Controllers who work the handoff positions need to keep the general picture of what is being said and done by the pilots and controllers who are working the associated sectors. These people are most effective when they can react to a pilot request that requires coordination without the radar controller having to repeat that request to them. By monitoring the position, they hear the request first hand and can be in the process of conducting the operation while the controller working the sector is responding to the pilot and doing other control functions. This capability is most important during emergency situations where time is critical and help is most often needed. The sector controller can concentrate on providing radar vectors, emergency airport information, and current weather conditions, while the handoff person gathers data, provides information requested by the pilot, or notifies the appropriate emergency equipment. Most supervisory consoles allow the capability of monitoring any position associated with that supervisor's area of responsibility. We use them to maintain an awareness of what is going on during these types of situations. We also use these positions to conduct evaluations, certify controllers in training on new positions, and generally try to maintain the overall picture of what is going on within our area of responsibility.

You might think that all of the information I have just given you is a lot of nice-to-know, gee-whiz information, but are wondering what good all of this to you, the pilot. The major point of this entire chapter is that the more you know about how the system works, the more likely you will be able to work within the system and have the system work for you. Let me give you a few examples of how pilots and controllers can become very creative and mix and match equipment capabilities to get the job done, or in some cases, get it done faster, cheaper, and safer.

Improvisation

At one facility where I was working we had issued a clearance, over the phone, to an aircraft that was departing a satellite airport. This MU-2 was going to fly through our airspace while climbing enroute to the ARTCC sector above us and then on to another small, but tower-controlled airport in an adjacent ATC facility's airspace. The aircraft was on an IFR flight plan, but the flight was being conducted in VFR conditions. Shortly after takeoff the pilot realized that while he could hear the controllers, he did not have the ability to transmit back to them. The pilot could have conducted this flight in accordance with lost communication procedures, but he used another tool at his disposal that is not normally associated with ATC procedures.

This aircraft, or one of the people on board, had a modular telephone, so the pilot placed a call to our facility on a commercial telephone circuit. By transferring the call to the supervisors desk and having a handoff controller call into the supervisor's console we now had a three-way relay of the pilot's replies to the control position working the aircraft. Since the pilot's destination airport was also its home base and maintenance facility, it was agreed that the flight would continue in VFR conditions (the pilot cancelled his IFR when he first called the facility) and at a lower altitude so that he would remain in a tower-enroute environment. We transferred communications to the next ATC facility by providing the pilot with the frequency on which to listen and the facility's telephone number so that he could reply at ATC advisories. The controllers at the next approach control facility arranged for a landing clearance and taxi authorization at the destination airport while the aircraft was still in the air and under their control. What could have been a real problem was resolved by some very creative adjustments on the part of the pilot, and the controllers and our two facilities received a very nice letter of appreciation from the aircraft owners. This is not what you would call standard ATC procedures, but one of the first paragraphs in our operating handbooks tells us to use our heads when the situation is not covered by standard procedures. So this is what you might call a thinking person's ATC procedure.

I was personally involved in another situation where the combination of two ATC facilities, telephone land lines and radios, worked together to help a pilot. I had just transferred radar identification and communication of a twin Beech cargo aircraft to an ATRCC sector one night on a midnight shift, when the controller called back to say that the pilot had not called, was now squawking emergency, and appeared to be circling. The controller advised that the pilot was not responding to his calls and asked if I could still talk to him. The aircraft was now outside of my radar coverage area so I could not see the target, but I called the aircraft and asked for a reply and an ident on the transponder. All that I could get back was a garbled, unreadable reply, but the Center controller said that he observed an ident so the pilot could obviously hear me. Over the next few minutes the Center controller gave me instructions to relay to the pilot for vectors to the nearest airport and he indicated that the pilot appeared to be following instructions and identing to indicate he had received those instructions. Between us, we vectored the aircraft over a small airport, and the Center controller observed the aircraft starting to circle the airport before he too lost radar contact.

While we were waiting for the pilot to call, the Center controller had his supervisor call the state police and ask them to dispatch a vehicle to the airport to check on the aircraft. When we later talked to the pilot, we learned that he had experienced engine difficulty and that he had used our instructions to locate a suitable landing site. While this was not exactly a life-threatening emergency situation, it does point out how we can use a combination of communication systems to provide that service.

We also see situations where pilots get involved in some creative communication techniques. We have had pilots call on the emergency frequency because they had lost communication capability with the sector controller, and they knew that someone would be monitoring the emergency frequency who would be able to find out whom the pilot should talk to next. We have had pilots relay communications from other aircraft who were too low to be within our radio coverage area or who were in a blind spot. Pilots who are very familiar with the placements of remote transmitter/receiver sites often extend the range of their aircraft in favorable low-altitude wind conditions by using this communication capability in their route planning. They can also reduce their downtime by planning IFR training exercises to airports with those capabilities and picking up their outbound IFR flight plan directly from the controller instead of going through flight service.

These are just a few of the examples of how pilots and controllers can get creative as a result of a need. We frequently extend the limits of equipment by rigging an imaginative series of land-line or radio configurations so that we can make the systems do things that they were not designed to do. This is usually in response to an emergency or to help expedite pilots or other controllers, and the resultant configuration often becomes a standard procedure once discovered. As an example, one of our controllers figured out a way to monitor a control position that supposedly could not be monitored from where he was sitting. He plugged in a handset to a handoff position next to that radar scope and selected this monitor function on that position. What was being said on the control position was not being received in the handset at the handoff position. By doing this at several positions and then using the override capability at his own position to listen in on those handoff positions, he could hear what was going through those handsets. He had essentially created a monitor capability at positions not designed for that function. We routinely use that capability today.

The above examples will give you some indication of how a thorough knowledge and understanding of equipment capabilities can be used to increase the cooperation level between pilots and controllers. When they are operating within their own element, most pilots and controllers are very competent and knowledgeable about the equipment associated with their own job. Additionally, some pilots and a fairly large number of controllers have a relatively sophisticated understanding of the requirements and the equipment capabilities associated with the other's job. But there is one piece of controller equipment that few pilots know anything about, and this lack of understanding probably causes the controller more problems than are associated with any other piece of equipment.

FLIGHT DATA SYSTEMS

We talked a little bit about the FDEP (flight data encoding printer) and FDIO (flight data input output) systems in Chapter 2. (The accompanying photos will give you some idea of what this equipment looks like.) I gave you some examples of how a communication breakdown could result from a pilot misunderstanding of how this works. As I have indicated before, most pilots don't know what kinds of flight plan information the controllers have at their disposal. The resulting ignorance is frequently a source of incorrect assumptions, erroneous expectations, and just plain mistakes. So now let's delve into the operation of this equipment somewhat more deeply and demonstrate some of the traps that can result from its misuse. Hopefully, this will provide information about some of the ways to avoid those traps.

Section 1		Section 2	Section 3	Section 4	Section 5		
1		5	8	9	10	11	12
2	2A	6			13	14	15
3							
4		7		9A	16	17	18

The discussion of flight progress strips in this chapter is keyed to this format. This format is used at terminal ATC locations. ARTCCs use a somewhat different format.

N274FH	5122	6AØ	+EA1 AHN ELW SPA CLT+			
PA31/R ATZD	P2ØØØ		6AØ TOC CLT			
986	9Ø					

EAL284	4175	ATL	↑NO1 CHA CHATRANS SYI SYI1			
DC9/A	P2Ø15		ATL CHA SYI1 BNA			
242	31Ø					

N146ØT	6362	FTY	+SO2 GRANT ABY+			
PA34/A	P2Ø15		FTY ABY V159 GEF V159 ORL			
Ø74	7Ø					

Proposed outboard flight progress strips.

DAL1497	7Ø35	A1415	IFR			
1 B737/A	MACEY					
148	LOGEN		ATL			

Inbound flight progress strip.

R18Ø67	5142	E1331	8Ø			
U21/R	ATL		AGS./.V18 TDG ANB			
274	ZCT					

Overflight flight progress strip. The handwritten V in Section 5 calls the Atlanta controller's attention to this overflight.

This is the last piece of equipment that we will cover in this chapter, but is by no means the least important. I would also point out that the material covered in this section is more relevant to the terminal or approach control environment than to the ARTCC. The center also uses paper strips similar to the ones I will describe here, but their format is slightly different and they do not look exactly the same. Additionally, each ARTCC host computer may have subtle differences in their programming relative to the area that they control. These disparities will cause slight differences in the information that is printed on the strips. In general though, the following information is fairly indicative of what kinds of information the controllers have at their disposal when we are talking about flight plan data.

Flight Progress Strips

The forms pictured in this section are called *flight progress strips*. They are samples of the flight plan information that the controller receives when a pilot files a flight plan or when a flight plan has been activated. I have included examples of the various types of flight progress strips including *proposed outbound, inbound,* and *overflight* strips. The two outbound and inbound strips on N123TA will relate to a scenario which I will shortly use to explain one of the difficulties of duplicate flight plans. The others are simply samples of actual flight progress strips that we receive every day. The inbound and overflight strips are, of course, only generated if a departure flight plan is activated.

Outbound strips are copies of an individual flight plan that are sent by the FAA's computer system to the ATC facility(s) that will control the departure of that aircraft. This could be an air traffic control tower and associated approach control, or an air route traffic control center. Similarly, the inbound strip is only sent to the FAA facility(s) that will control the aircraft into the destination airport. I use the plural because some departure points or destinations may be VFR towers that have FDEP capability, and they will receive the same data as the approach control facility or Center that works the aircraft into or out of that airport. Any other ATC facility that works this aircraft enroute between the departure and arrival points will receive some form of an overflight strip.

Let me explain these flight progress strips by dividing them into sections and describing what each element means. (Refer to the number-keyed format and sample strips accompanying this discussion.) I'll go into details on some specific elements, try to explain the concept of the strips and the relationships between some of the individual elements, and give you examples of where the system breaks down when some action is required around a particular element. These flight progress strips are normally limited to IFR flight plans, but as we will shortly discuss, they can be used in a VFR environment.

Section 1. The first section contains the aircraft call sign (Item 1), the aircraft type with equipment suffix (Item 3), and a computer identification number (Item 4) assigned specifically to this flight plan that is for air traffic use only. Equipment suffix designators vary with the type of equipment in the aircraft. An R suffix, for example, indicates that the aircraft has area or inertial navigation equipment. Occasionally there will be a subscript number between the call sign and type aircraft similar to the "1" on the DAL1497 outbound strip (Item 2). This number indicates that an amendment was made to the flight plan by someone in the ATC system. The change could be to any element on the flight plan and a cross check between the two strips will usually make the change apparent. For this reason, controllers always use the strip with the highest amendment number to ensure that the most current information is being used. There may also be a series of numbers and/or letters just to the right of this amendment number similar to the "ATZD" on the departure strips from 6A0 (Item 2A). This indicates that someone requested a copy of this strip or forced a copy to the receiving facility. These alphanumeric characters indicate which sector or facility initiated that action. The information in this section is generally the same for all types of flight progress strips.

Section 2. The second section contains three elements, two of which will contain different information depending on what type of flight progress strip is in use.

The top element, which is the same for all types of flight plans, is the computer-assigned transponder code (Item 5). When an aircraft is not equipped with a transponder or when the pilot fails to include the equipment suffix with the flight plan information, this section is left blank. It is amazing how many flight plans we get that show sophisticated aircraft without transponders. The FSS specialist will not try to guess what kinds of equipment an aircraft has on board, so if a pilot drops off a flight plan that shows no equipment suffix, that's exactly what the specialist types in the computer. I doubt that there are too many Learjets running around without transponders, and each time this happens we have to wait until we can talk to the pilot before we can make the amendment. If the aircraft is in the air, the usual result is a delay to the pilot.

In a proposed departure flight strip, Item 6 refers to the proposed departure time that the pilot (or in some cases the company operations/scheduling office) has given to the FSS when the flight plan was filed. This information is always in Zulu time, so don't make an all-too-typical pilot error and give local time to the FSS. If you do, you will end up with a flight plan that has either timed out (we only keep them in the system for two hours after the proposed departure time) or one that won't show up in the computer until tomorrow. If you file a flight plan for 1445 thinking 2:45 P.M. and it is already 1545Z (10:45 A.M. local

Atlanta time) when you file it, the flight plan will enter the computer at 1445Z—23 hours later. Most FSS specialists will ask you about this if it is questionable, but sometimes they do not get the chance to do so.

Item 7 in an outbound flight strip refers to the altitude requested by the pilot for this flight plan. We rarely see mistakes in this item, and when we do, it is usually quite obvious. (Something like a Learjet requesting 2000 instead of 20,000.) By typing the contraction VFR into this element, a controller or FSS specialist can generate a *VFR flight progress strip* similar to the IFR strip that we have been discussing. The only difference will be the VFR in Item 7 of the outbound strip and a VFR contraction in Item 9 of the inbound or overflight strip. The biggest problem we have, related to Item 7, is that some pilots have a tendency to change their altitude request after they are airborne and we have completed the handoff to the next facility. Let me explain the problems that this causes.

Most approach control facilities are assigned airspace from the surface to 10,000 feet. Atlanta is an exception to this in that we own up to 14,000 feet. Some of the smaller facilities only own up to 6000 feet. When a pilot files a flight plan between two approach control facilities that have a common boundary and requests an altitude that is, in both cases, within their assigned airspace, two things happen. The ARTCC host computer loads the information into the local facilities' ARTS and/or FDEP/FDIO memories, and the overlying center sectors know nothing about this flight. If the pilot then waits until the approach control facility that controlled him initially has completed an automated handoff to the next approach control facility before requesting an altitude change that will take the aircraft into ARTCC airspace, we have the following scenario. Let's make the first tower facility A, the second facility B, and the ARTCC facility C.

The aircraft will be in facility A's airspace and usually still on their frequency. Facility B will have control of the ARTS track and computer control of the flight plan information. Any attempt by facility A to amend any of the data on the flight plan will result in a computer message that says something to the effect of, "Look dummy, you already gave away this aircraft. Not your control."

Facility B can't climb the aircraft because, first, they are not talking to the aircraft and second, they own neither the airspace that the aircraft is in nor the airspace to which the pilot wants to climb. Facility B also cannot initiate a handoff to facility C without making an altitude amendment in the flight plan because the computer knows that the current altitude is not in the ARTCC's airspace and it won't permit such a handoff.

The controller at facility C knows absolutely nothing about this flight and if he or she is smart (there aren't too many dumb center controllers) will say, "Let me know when you get all the paperwork straightened out and send me a copy of the flight plan."

So, what is required to change an altitude into, or for that matter, out of the ARTCC structure (the situation is just as bad when the pilot wants a lower altitude) is as follows.

The controller at facility A has to call the controller at facility B and explain the situation and ask them to make an amendment into the FDEP/FDIO. The controller at facility B has to make the amendment and call facility A back to let them know that the amendment has been made. They also initiate an automated handoff to facility C which gives them track and computer control. Facility A then has to call facility C and explain the situation, probably wait while the controller obtains the flight plan that is just now printing out on their equipment, and make a manual handoff to that controller. Finally, the controller at facility A has to go back to the pilot and assign the new altitude.

While all of this was going on, the pilot's spouse has decided that flying along just over the tops of the clouds is not really all that bumpy and is absolutely beautiful, so they now want to stay at 10,000 feet. It is at this point that we launch the ground-to-air missiles. Let's get back to our discussion.

On an inbound flight strip, Item 6 contains what we call a *coordination fix*. The receiving controller can look at this information and know a point in the preceding controller's airspace over which the aircraft will be routed. This information combined with the data in Item 7 of an inbound flight strip, which is referred to as the *transfer of control/communication fix* and is usually a point within the receiving controller's airspace, will give the receiving controller a straight line route of flight between these two points along which the aircraft will be flying.

Most facilities have internal procedures that determine which controller works aircraft over specific inbound routes. So, if you are ever instructed to contact a different facility and that controller seems not to know anything about you, either you or the information about you has been given to the wrong controller. Normally, in this case, the controller will ask you where you are and where you are going. He or she will then look at the aircraft that would normally be controlled by that sector to see if your track is there and assign the correct frequency. If they cannot find you or it is obvious that you are talking to the wrong facility, they will send you back to the previously assigned frequency to check where you are supposed to be. If you are one of those pilots who doesn't write down the list of frequencies that you have been on, uses the same radio for every frequency change, and has a short memory—good luck.

Section 3. The third section normally only contains one item. On an outbound flight plan, Item 8 is the departure airport or point in space where the flight plan will begin. ATC facilities which control traffic off of several airports will normally group those flight plans with the same departure point together. This is so that they may be located quickly from among the dozens of other flight plans when the pilot calls for clearance.

On an inbound or an overflight flight plan, Item 8 is a Coordinated Universal Time (UTC or ''Zulu'') figure. This figure estimates when the aircraft will be at some point on its route, usually the coordination fix, that is relative to the facility receiving the flight plan. Its uses will become more apparent when we explain the elements of an inbound strip in section four.

Section 4. The fourth section contains several different types of information, and this information may be located in different areas within this section. A lot depends on how much information is contained in the actual flight plan and whether it is an inbound or outbound flight plan. Let's talk about an outbound flight plan first.

In an outbound flight plan, Item 9 will be the route of flight that the pilot has filed unless that route has been changed for some reason. There are several reasons why this change could be brought about but the most common reasons are that the filed route was incorrect. Other reasons include the possibility that a navigation outage forced the change or the departure or destination airports triggered a *preferential departure route* (PDR) in the computer.

PDRs are established to regulate and align traffic into or out of high-density airports and route structures. PDRs are frequently assigned whether you file that route or not because the computer will automatically assign the correct route for that segment of your flight plan. Pilot's can, of course, request a different route, but these PDRs (see ''Preferred IFR routes'' in the *AIM* Pilot/Controller Glossary or the *Airport/Facility Directory*) are set up for specific route/altitude segments, and you may have to change altitudes and/or obtain the authorization from the individual ATC facility or sector to fly a conflicting route of flight.

(Assuming that a route change has been forced into the flight plan, the information regarding this change will now occupy the top part of Item 9 in an outbound flight plan and the filed route will move down one line. The change will be bracketed by symbols. In the case of a PDR, there will be a plus sign before and after the new route. (Up and down arrows indicate a *preferential departure and arrival route*. This is a route between two terminals that are within or adjacent to one ARTCC's airspace.)

A controller will issue your clearance as follows:

Pilot: *XYZ Tower, N123TA request IFR clearance to Medianville.*

Controller: N123TA, XYZ tower, you are cleared to the Medianville airport as filed, except change route to read ...

The ''except change route to read'' segment will be the information contained within the pluses/arrows on the controller's flight plan information. This new flight path will either be an entirely different route from departure point to destination or it will take you from some point on your filed route to some other point on your filed route.

In the absence of a change to the flight plan, the filed route alone will occupy Item 9. But in either case, if the filed route is longer than the space available to type this information, the computer will shorten the data by inserting some kind of flag into the flight plan. This announces that more data is available on this route, but you must ask the computer for a full printout. This situation is referred to as *truncating* or shortening the flight plan, and it would look something like: ATL V97 ABY ./. MIA. We must ask the computer for the information about the route of flight between Albany and Miami.

On a departure strip, Item 9A in this section contains information concerning any remarks that are attached to the flight plan. It is in this location that the computer or the FSS specialists insert the contraction FRC (full route clearance) that alerts the controller that some element of the flight plan has been changed from what the pilot filed. This places a requirement on the controller to read the entire flight plan to the pilot so that the pilot becomes aware of any changes. Most controllers will tell you why they are reading you all of this information, but we are not required to do that. Additionally, when a controller reads an FRC, he or she may just read the change into the flight plan at the proper location which eliminates the requirement to say "change route to read." You are, of course, expected to be listening when you ask for your clearance.

Other information that you may have asked to be placed in this section will also be included up to the limits of the space available. If there is more information than there is room, the presence of additional information will be indicated by a series of stars at the end of Item 9A. Again, this information can be obtained by asking the computer for a full readout of the flight plan.

Inbound and overflight strips contain slightly different types of information in Section 4. Inbound strips contain either the contraction VFR or IFR in Item 9 while an overflight strip will show an altitude in this area. You might wonder why there would be no altitude information on inbound strips, but the reason is relatively simple. Most facilities use a structured arrival procedure where aircraft inbound over a particular route will be assigned a particular altitude. Any exceptions to this prearranged procedure have to be coordinated in advance so the controller always knows what the altitude of the inbound aircraft will be.

Since we always verify an aircraft's altitude on initial contact and have a Mode C altitude readout on many of the aircraft we work, there is little likelihood that we will be surprised by an unexpected procedure. Overflights, on the other hand, have a wide range of routes and altitudes on which they can operate through a facility's airspace, so the altitude information on a flight progress strip is a method of advance coordination.

The second line of Item 9 on an inbound strip is usually blank while this same area on an overflight strip contains the route of flight on which the aircraft will be operating. This route may be truncated if the whole route will not fit in the space, but the section of the route that applies to the facility receiving the strip will normally be printed in as much detail as possible.

On an inbound strip, Item 9A contains the destination airport and as much of the remarks information as space will allow. Again, the stars or asterisks at the end of this information indicates that more data remains than could be printed. A controller would need to request the information from the computer.

Section 5. The fifth section of the flight progress strip contains nine items. This section is reserved for local strip writing procedures and usually does not contain any preprinted data. Local or national flow control programs which generate *estimated departure clearance times* (EDCTs) are the exception in this section. Let's discuss this program in more detail.

EDCT

We have talked about flow control programs on several occasions in this book, so let me take a little time while we are on the subject to elaborate on one of aviation's favorite topics, ATC delays. EDCT programs are computer-assigned departure times that regulate the flow of traffic into a busy airport. These times are based on the speed of the aircraft and the estimated time that a given aircraft would arrive in the destination airport's traffic picture. If there are too many aircraft scheduled into an airport within a given time period, the computer will randomly select the correct number of flight plans and adjust their departure times so that they will arrive in the receiving airport's traffic picture during a slack period.

There are only a few airports in the country that have a heavy enough arrival schedule to routinely require this controlled departure program, but these methods can also be used to adjust to unusual circumstances. For example, Atlanta can theoretically accept 105 arrivals per hour under ideal circumstances. Those conditions would call for simultaneous visual approaches with four operational runways and no weather which could impact the operation.

It is very rare that there are more than this number of scheduled arrivals during a given hour, so when these conditions exist there are usually no EDCT programs in effect for Atlanta. If, however, we scatter a few thunderstorms around the area, close one of the runways and have to conduct simultaneous ILS approaches because of poor visibility, our acceptance rate drops to about 75 an hour. If there are 80 aircraft scheduled into the system for that period, 5 of those aircraft will be selected by the computer and given an EDCT time that moves their arrival time into a less crowded time slot. From this you can begin to see why some airports encounter delays during severe weather situations or extremely busy travel periods such as the Christmas holidays.

EDCT programs can also be used to regulate traffic during unusual circumstances or major events that draw large numbers of aircraft to an otherwise low-traffic area. Typical of these would be a place like Louisville during the Kentucky Derby or Indianapolis during the Indy 500. If my memory serves me correctly, we also had some controlled departure programs into Cincinnati for a few days after a tornado nearly wiped out the control tower on that airport. Regardless of the reason, these types of regulated flow control programs are designed to keep the system full of aircraft but not overloaded, and subsequently reduce the safety margin, at those facilities that could easily be inundated by "just a few more airplanes."

These programs impact all aviation segments, and each of these groups has learned how to work within the system in order to get around the programs. Airlines will work with the flow controllers to swap delay times for a critical flight. They will take a longer delay time for aircraft number 2 off of airport B, so that aircraft number 1 off of airport A can use that EDCT time to meet the connections for its passengers. The 90 passengers on aircraft number 2, which was a DC-9, may then be combined with the 90 people on another DC-9, scheduled 30 minutes later and going to the same airport. This larger group of people would then be placed on a wide-bodied jet capable of handling them all. This technique creates an extra slot by eliminating one flight and the result is that neither of the aircraft receive a delay that will cause their passengers to miss their connections. There is, of course, a limit to the number of aircraft that a given airline has at its disposal to jockey around, and eventually some people are going to be delayed. But the airlines use this technique to the maximum advantage.

Similarly, the corporate and general aviation segments have learned to adjust their scheduling programs when they have to travel to areas where the traffic frequently overwhelms the concrete available for landing. Many of them will schedule their trips in the evening and spend the night before an important meeting in a hotel rather than the night after that meeting. They frequently find that their people are more rested and effective in this environment. They might even reluctantly agree that the net result was usually the same even though they had to adjust their schedules.

These segments of aviation have also discovered the joys of reliever airports. By filing into a smaller field that is 20 to 30 miles from a major terminal, they can schedule their departure times whenever they want. If they give themselves an extra half hour and arrange for ground transportation, they can easily make their meeting. They usually have to have transportation anyway, and a limo will really impress the client. The company can probably absorb the cost of the ground transportation from the savings in fuel that they will realize by not having to go into a holding pattern near the very busy airport.

Most good chief pilots and scheduling offices learn very quickly how to work within the system to get their job done with the minimum impact on their company and their bosses' schedule. Unfortunately, there are situations where even the best laid plans go awry and we have to deal with irate pilots who are in a hurry and going nowhere. In some cases, the paperwork can lag behind the creative minds of the people doing the shuffling, and the ATC flight data specialist sometimes has to explain to the pilot that they have not yet received their flight plans. In these circumstances, the controller usually gets on the phone and starts coordinating to expedite the operation. So try to exercise a little patience and know that we are trying to get you on your way. You might say that we want to get rid of you as much as you want to get away from us.

Some pilots even try to get around the system by launching off of the airport VFR and trying to pick up the flight plan in the air. This doesn't work for several reasons. First, there is a high probability that their VFR departure will have resulted in their IFR flight plan being canceled. The controller at their departure airport will normally ask the pilot to verify that they wanted to depart VFR to their destination airport. If an affirmative reply was given, they remove the IFR flight plan from the computer.

Second, no air traffic controller is going to activate a flight plan with a delay time without first checking to see if the flow controller and destination airport can absorb that traffic. Generally the answer is going to be no, and this pilot is now stuck in the air without an IFR flight plan. This type of pilot will then probably climb to the highest VFR altitude that the regulations allow, do so without ATC advisories, and cruise through and over airspace that is saturated with an aluminum overcast.

When they get to their destination area, they will act like a VFR pop-up and try to talk their way into the system at that point. Even though these pilots are trying to beat the system and would probably just as soon work against us as with us, we still try to accommodate them whenever we can do so without compromising the safety of other aircraft. Although this tends to reinforce the fact that they can get away with what they want most of the time, our job is to move the airplanes and not police their morals and ethics. Sooner or later this type of pilot runs into a situation where they violate some FAR and get caught. Our only hope is that a "situation" is the first thing that they run into.

You now know most of the same information about a flight plan that a controller knows when one of these paper strips is delivered to a sector. Most facilities use some form of internal strip marking to highlight specific information on these data strips. Atlanta uses a large V in Section 5 to call attention to an overflight strip. We also use some symbol in Section 3 of arrival strips to indicate over which fix an aircraft will arrive. But beyond these enhancements and other locally adapted strip marking procedures, what you see on these strips is all we know about a flight plan when you arrive on the frequency or call for clearance. So if you have just talked to flight service about a change to your flight plan or are departing early, let's not make this information your little secret.

We talked at some length in Chapter 2 about the problems caused by the lack of a flight plan in the system when you call for clearance. So there is really no need to elaborate further, other than to ask you to combine that knowledge with what you have just learned is this chapter. Do your best to make sure that the ATC system is ready for you when you are ready for them.

Double Your Flight Plan, Double Your Fun

Now that we have forever solved the problems of the missing flight plan, let's talk about the other end of that same equation, the presence of more than one flight plan for the same aircraft. You should know that there is just as much potential for confusion and problems in this circumstance as there is when no flight plan exists.

There are as many reasons for why this situation might occur as there are aircraft, but other than the occasional mistake, the usual reason is that a pilot is trying to save time and prepare for any circumstance.

Let's construct a scenario that almost any corporate pilot will recognize. In this instance, we will assume that a company's CEO is in Macon for a meeting, and that the outcome of the meeting will determine what the boss does next. If the meeting goes well, the next move may be to return to the corporate offices in northeast Atlanta. Since the company has a hangar at the Dekalb Peachtree Airport, they would return to PDK. If the outcome of the meeting requires the boss to travel to their international offices in Europe, the pilot would have to deliver the CEO to the Hartsfield airport to catch a flight.

Typically, the pilot covers both possibilities by filing two flight plans. (For this scenario, refer to the outbound and inbound flight progress strips for N123TA.) As long as the proposed departure times are similar, Macon Tower will receive the paper documents for both flight plans. However, they will not know which one went into the computer first unless they check to see which flight plan is

N123TA	2607	MCN	MCN TRBOW TUCKR PDK			
1 BE20/R ATZD	P1445					
622	110		O TEST FP NO SIDS NO STARS			

N123TA	5245	MCN	↑V35 SINCA↓	EDCT 1530		
BE20/R ATZD	P1500		MCN ATL			
139	110		O TEST FP FRC NO SIDS N***			

Proposed outbound strips for N123TA.

N123TA	2607	A1458	IFR		
BE20/R	MCN				
622	TRBOW		PDK	O TEST FP NO SIDS***	

N123TA	5245	A1457	IFR		
BE20/R	SINCA				
139	HUSKY		ATL	O TEST FP FRC NO ***	

Inbound strips for N123TA.

in their ARTS tab lists (assuming, of course, that their facility is equipped with ARTS capability). At the time I was writing this book, Macon Approach was equipped with ARTS, but Columbus Approach, another facility that has a common boundary with Atlanta and could easily have fit into this scenario, did not. Had I used Columbus in this hypothetical situation, they would have no way of knowing which flight plan was filed first.

The ARTS system will not normally allow two or more flight plans with the same call sign to be stored in the active tab list associated with a radar scope. If the pilot wants to use the one that was filed second (isn't that always the case), Macon would have to remove the first flight plan from the system and force the second one into the ARTS. This takes time and the Big Boss is always in a hurry. If, for whatever reason, the wrong flight plan is activated, things could get interesting (remember "Top Gun" and J.R.?).

Let's now walk through a possible set of circumstances that will lead the aircraft into the wrong airspace, at the wrong time, destined for the wrong airport and show you how easily this can happen if everyone is not on their toes.

Note that N123TA's flight plan to PDK is scheduled for 15 minutes earlier than the one to ATL (P1445 vs. P1500). There are any number of reasons why a pilot might file this way, but for our purposes we will just accept that fact. The result will be that the PDK flight plan will be delivered to the clearance delivery person 15 minutes before the one to ATL. Now let's say that a conscientious copilot, who knows that the aircraft is on file to Atlanta but knows nothing about the contingency flight plan, calls for the clearance sometime during that 15 minute period. The following set of transmissions might possibly occur:

Pilot: *Macon Tower, N123TA, on the general aviation ramp, ready to copy IFR clearance.*

The controller locates the only flight plan for N123TA that currently exists and studies the information. The PDR into the Peachtree Airport calls for all aircraft to be vectored over the TRBOW intersection and then via the TRBOW STAR into PDK. These pilots have requested no SIDs or STARs in their flight plan so the controller would usually have to issue a complete readout that covers all of the fixes in the STAR route. Normally that would involve telling pilots that they were cleared to PDK via radar vectors to TRBOW, direct TUCKR, direct PDK. But if you will look at the flight plan, you will see that this is exactly how the pilot filed. Additionally, the crossing altitude at TRBOW is 11,000 feet, which is also exactly what the pilot is requesting. This is not as uncommon as it might sound, and because of this, the controller is not surprised and issues the following clearance:

Controller: N123TA, Macon Tower, you are cleared to the PDK airport as filed, maintain 3000, expect 11,000 within ten minutes, departure control frequency 124.2 squawk 2607.

Now let us make a couple of assumptions that are not as far fetched as anyone might think, and we will have begun to set the stage for the fire drill that will occur 45 miles southeast of Atlanta a few minutes after this aircraft departs. First, we will suggest that the copilot did not hear the word PDK in the clearance. Remember, we often hear what we expect to hear and not what was actually said. This is especially true if the copilot was trying to do several things at once, which is not unusual. This pilot has written the information about the clearance on his/her scratch pad using the contraction CAF (cleared as filed) to cover the route segment of the flight plan and has all of the rest of the data correct.

Now we will eliminate a possible method of catching the error by giving the Macon Tower clearance delivery controller a coffee break. This individual is relieved by another controller within this 15-minute buffer between the printing of the two flight plans and after the first controller has read a clearance to N3TA. The relieving controller is not familiar with the first flight plan which is now mixed with several other flight plans in front of the ground controller, and the controller being relieved has no reason to make that person aware of the situation. When the second flight plan (the one to Atlanta) for N123TA prints out, the clearance delivery controller will add it to the stack of other unread flight plans and wait (a *long* time in this case) for the pilot to call for clearance.

When N123TA taxis for takeoff, there will be no reason to delay the aircraft, no unusual circumstances which would trigger a question in the mind of the pilots or the controller. After the aircraft had been switched to departure control, the flight plan to PDK would have been automatically activated because the pilot was broadcasting the assigned beacon code. Since this was the first flight plan assigned to the ARTS, the data was also in the controller's tab lists. The Macon departure controller assigns the aircraft a heading that will aim the aircraft toward the TRBOW intersection and climbs N123TA to Macon's highest usable altitude which is 10,000 feet. There are few clues that something might be amiss in these actions, because the heading to TRBOW is only a few degrees east of a heading that would be assigned to an aircraft being vectored toward the Atlanta arrival radial, and the altitude assignments will be the same.

The Macon controller now initiates an automated handoff to Atlanta ARTCC (they own 11,000 feet over the Macon/TRBOW/Atlanta arrival area) and instructs the aircraft to contact Atlanta Center. Again no clues, because these same procedures would apply if the aircraft were being vectored to the Atlanta arrival radial.

The Atlanta Center controller observes that the heading assigned by Macon is taking the aircraft almost direct to the TRBOW intersection. This concurs with the overflight strip that this controller has, so he or she only verifies the aircraft's Mode C altitude readout and climbs N123TA to 11,000 feet. Once more, there is no unusual activity to trigger a question in the minds of the pilots. The ARTCC controller then initiates an automated handoff to the Atlanta controller who is working the southeast satellite sector and controlling aircraft inbound to airports like PDK. The ATL controller notes that the heading is right for an intercept of the route over TRBOW, the altitude is correct for the airspace owned by that sector, and the inbound flight plan to PDK concurs with all of the above. The controller takes the handoff.

The Center controller has now completed almost all of the work required for this flight and issues the following transmission:

Controller: N123TA proceed direct TRBOW, flight plan route, and contact Atlanta approach control on frequency 119.8.

It is at this point that the bells go off in the pilots' heads. They inform the Center controller that something is wrong and the fire drill begins.

Had the controllers known this aircraft was going to Atlanta, it would still be on the ground at Macon. There were EDCT programs in effect, and this aircraft would have had to absorb a 30-minute delay because of traffic saturation. Right now though, this aircraft is just outside of Atlanta's airspace, the traffic saturation into the big airport still exists, the aircraft is on the frequency of a controller who no longer has control of the flight plan, it is headed for the airspace of a controller who knows nothing about the problem that just arose, and the aircraft is doing all of this at 210 to 250 knots. Someone will have to absorb this aircraft and separate it from everyone else for as long as it takes to get this mess straightened out. Some controller will have to coordinate all of the new procedures, frequencies, and flight data information and create an arrival slot for one unplanned aircraft—all of this because one pilot missed one word in one clearance.

I hope that you now understand why it is so important that every pilot know as much as possible about the capabilities and limitations of the equipment that controllers and pilots use in the everyday aviation environment.

5

Regulations and Procedures

IT HAS BEEN SAID THAT READING FEDERAL AVIATION REGULATIONS (FARs) AND air traffic procedures handbooks is something akin to studying a dictionary. They are written in the most difficult of all languages to understand, legalize, and if stacked end to end would take up more space than the first flight at Kitty Hawk. Compare today's regulations against the list of the original FARs on the next page of this chapter and you can see that we have come a long way. Or have we?

All kidding aside though, the first FARs were simply a list of common-sense procedures that laid out in black and white what most pilots did as a matter of practice. It is entirely possible that some of them, such as the one about the spurs, came about as a result of one of the lessons learned from an accident. This is still true today, and if there is a link between today's procedural rules and those early guidelines, it is that they were designed to enhance the safety of flight. Then or now, if they are followed and understood by all concerned, FARs form the basis for one of the safest and most efficient aviation systems in the world.

Unfortunately, regulations and procedures are probably one of the weakest areas for the average pilot and air traffic controller. Aside from the sheer volume of information, there are a number of other reasons why our two groups are failing to keep abreast of each other's procedural requirements.

First, the actions of our two groups are governed by two completely separate sets of regulatory documents. This is, of course, intentional, because much of the information required by one group is not needed by the other. The average pilot doesn't care how we set up our radar scopes as long as we do it right. Most controllers are not concerned about what kind of fuel you put in the aircraft as long as it keeps running while it's in the air. But, intentional or not, the net result

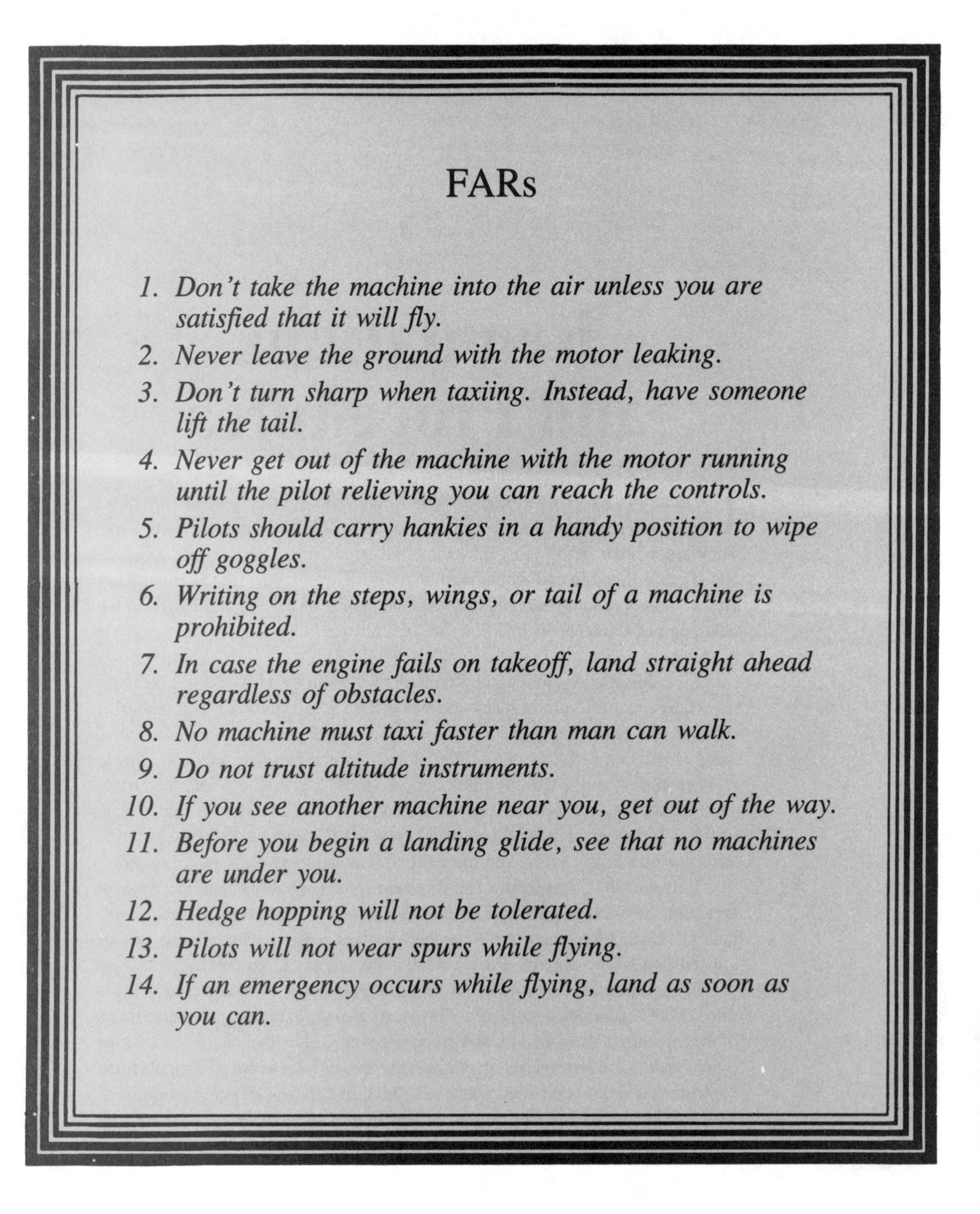

FARs

1. *Don't take the machine into the air unless you are satisfied that it will fly.*
2. *Never leave the ground with the motor leaking.*
3. *Don't turn sharp when taxiing. Instead, have someone lift the tail.*
4. *Never get out of the machine with the motor running until the pilot relieving you can reach the controls.*
5. *Pilots should carry hankies in a handy position to wipe off goggles.*
6. *Writing on the steps, wings, or tail of a machine is prohibited.*
7. *In case the engine fails on takeoff, land straight ahead regardless of obstacles.*
8. *No machine must taxi faster than man can walk.*
9. *Do not trust altitude instruments.*
10. *If you see another machine near you, get out of the way.*
11. *Before you begin a landing glide, see that no machines are under you.*
12. *Hedge hopping will not be tolerated.*
13. *Pilots will not wear spurs while flying.*
14. *If an emergency occurs while flying, land as soon as you can.*

is that a large amount of material that every pilot and controller should know about the other person's job is not contained within the documents which govern our respective groups.

That leads us into the second reason why our knowledge of each others requirements is so weak. With the exception of those few cases where the rules or procedures overlap, there is no real requirement for one aviation segment to be aware of, or knowledgeable about, the regulatory procedures of the other. I am not trying to suggest that all pilots should be rated controllers and all controllers should be flight instructors, although this certainly would help. The point that needs to be made is that each group should be more understanding of the requirements of the other.

Every year I talk to dozens of pilots who corner me at pilot/controller forums or call to complain about controller actions which, when I investigate the situation, turn out to be not only proper, but required by ATC procedures. Similarly, controllers often fuss about why a pilot will conduct a particular operation or make an unusual request. When I review the situation, I have to point out that they, unlike many other pilots, are complying with some obscure FAR.

Beyond the fact that some members of each group lack a fundamental understanding of the other's procedures, there is a third development in aviation regulations that is perhaps the most difficult of all the areas we have to deal with. Over the years, the differences between what is a regulation and what is simply a good operating practice have blended together to the point that a very large percentage of both pilots and controllers do not know where one begins and the other ends. The distinction between the two often leads to some first-class disagreements between our two groups. In some cases, one or the other of us will make an arbitrary decision based on that misunderstanding—a decision that does not usually set well with the other.

As we continue in this chapter, I will cover as many of these misunderstandings as possible and tie them to the regulations which spawned the misconception. But, as a means of establishing a better understanding of how our system works, I want to talk first, in general terms, about the structure of regulatory and monitoring agencies on both sides of the radar display.

Just so you won't forget that this is primarily a chapter about FARs, let's preface the discussion by starting you off with a couple of little FAR teasers that can rattle around in your brain as you read on.

- First, let's assume that you are a pilot who is going to file an IFR flight plan from a small airport in Georgia. You are piloting a single-engine aircraft on a westbound flight. You also want to take advantage of favorable winds aloft that exist at 8000 and 9000 feet. What altitude will you file, and why did you choose that altitude?

- Second, you are now airborne on that flight. You have been radar identified, are being worked by a radar approach control facility on a Victor airway, and are climbing to your requested altitude. You are headed toward a destination where you will enter a holding pattern and then conduct an ILS approach. Consider all of the actions that will occur on this flight and answer the following question: What information are you required to give to the controller and when must this information be given?

Think on those for a while and remember that there are FARs covering both of these situations.

STAYING CURRENT

I certainly don't want you to get the idea that the aviation system is falling apart at the seams as result of poor teaching and lousy regulations, because it is not. Like most of the rest of our society, the organizations which are charged with teaching and regulating the various elements of aviation are being overwhelmed by the sheer numbers. More people and products have flooded the aviation market than the agencies can possibly handle. Also, the rapidity with which changes are occurring in aviation technology and law requires that the people in these agencies spend an inordinate amount of time just trying to stay current with change.

Is this an irreversible trend, with no help on the horizon? Certainly not, the situation is mostly a result of a lack of awareness on the part of most pilots about the knowledge levels that they should have at a particular point in their flying experience. Since they have already obtained a pilot's certificate, they think they have learned most of the things that they need to know to become a good pilot. This is not necessarily arrogance on their part, because most pilots know that their skills are weak. They simply believe that the only thing they lack in their goal to become a better pilot is practice. Isn't that what the flight instructor and the examiner have continually been telling them?

This is a case where they don't know that the learning process needs to continue after they obtain their certificate. But even after we convince them of this, they will still not be sure of where to obtain the additional knowledge. So, if we can create that awareness in the minds of these pilots and then give them a few resources from which to learn, the rest is relatively easy.

There is no shortage of organizations, publications, and individuals who will go out of their way to provide a full range of information to any individual who is willing to listen, read, and learn. If we establish some guidelines for the average pilot to follow, and communicate a general understanding of how the system works, they will usually educate themselves. With this in mind, let me give you a short overview of some of the techniques that are used to monitor the aviation system on both sides of the radar scope. I would also remind you again that you, the individual pilot or controller, are the one who is responsible for learning the material that will make you the kind of professional that you should want to be. All of the materials that are made available to you, including what small impact this book might have, are useless unless you take advantage of them and refuse to accept less than the best possible instruction for the money you spend.

QUALITY CONTROL

The FAA has had a long-standing commitment within the aviation community toward improving the quality of aviation in general and the skill level of the people who operate in the air traffic environment in particular. This commitment is not limited to any one element of aviation. Rather, the agency, in cooperation with airlines, aircraft builders, flight training facilities, corporate aviation concerns, various aviation publications and pilot organizations, private air traffic control organizations, and the military, is constantly trying to standardize products, procedures, training, and general knowledge.

Virtually every element of the FAA has some part of its structure dedicated toward a particular segment of aviation monitoring, education, and/or development. Some are charged with developing or monitoring procedures, handbooks, phraseologies, etc. that are used in every conceivable type of interaction between and among pilots and controllers. Others are charged with the responsibility of monitoring maintenance, crew currency requirements, training procedures, and so on. There is virtually no aspect of aviation that does not have some civilian and/or governmental group assigned as a watchdog or a resource point. Let's look at some specific examples of how this is done and how it affects each person in the air traffic environment.

FAA Evaluation Branch

Each FAA region has an organization within its structure called the *Evaluation Branch*. Nationwide, these branches consist of several hundred men and women who have the responsibility of monitoring the performance of the air traffic control system. They evaluate ATC facilities to ensure that they are conducting their operations in accordance with prescribed handbooks and procedures. They check the training programs to verify that each ATC specialist receives the proper amount of training and that this training is of a high quality. They monitor the actual operation of the ATC system to ensure that specialists are controlling air traffic in accordance with the proper procedures. They also review each instance of a reported near-miss and every case where less-than-standard separation occurred between two or more aircraft or between an aircraft and another controller's airspace. We will talk more about separation later in this chapter, but I included it here because it gives you an idea of the scope of the Evaluation Branch's job. In general, these specialists go over each facility with a fine tooth comb to make sure that the i's are dotted and the t's are crossed.

Evaluation specialists are not just limited to ground-based inspections and paperwork reviews. Each time that they fly from one point to another, either as pilots or as observers on the flight decks of commercial aircraft, they conduct what are called *airborne evaluations*. They take notes regarding the quality of the overall aviation system, and these notes, commenting on both good and bad observations, are then passed to the appropriate facility or organization for further action. During these evaluations they are not normally monitoring individual pilot performance but, if they observe a flagrant violation of FARs or procedures on the part of a pilot, they will pass that information along to the appropriate air carrier or general aviation flight inspector for investigation and possible action.

All of the information gathered from all of these evaluations is then analyzed and sent to the organization that is best equipped to act on the findings. If the information relates to a particular ATC facility, it is given to the manager of that facility. From there, it is usually given to the office of the facility quality assurance specialist(s) for action to correct any deficiencies or disseminate particularly good procedures to other facilities. This office usually consists of one or two persons who perform, within the facility on a continuing basis, many of the same tasks that the evaluation branch specialists do on a random or scheduled basis. Together, these two groups provide supervisors with continuing feedback on how the individual controllers on their teams are performing with respect to the proper use of phraseology, ATC procedures, and services to the flying public.

Controller Participation

The system I described works well as a source of information and a monitoring process, but unless each individual participates in the review of his own skills, the education process begins to break down and new information is never learned. Controllers participate in this process, both directly and indirectly, in several ways. Each controller must listen to a tape of their performance during actual traffic conditions at least twice a year. This is done so that the controller can listen to what they are actually saying, not what they think they are saying. They are also responsible for maintaining currency on each position of operation and on specific techniques, such as surveillance approaches, that are infrequently used but critical operations.

As part of the controller's review process, every supervisor directly monitors each controller on their team on a continuing basis to review the controller's adherence to phraseology and procedural standards. These supervisors also conduct at least two complete reviews every year with each specialist on their team covering every aspect of the controller's job. These reviews are followed by a debriefing session between controller and supervisor during which the quality of the work being performed is discussed and changes, if necessary, are agreed upon. Ideally, these discussions form the basis of a "contract" between the controller and the supervisor for the type of work that will be performed during their performance appraisal period.

Most facilities also have a monthly training period set aside for each team of controllers so that they can review various segments of ATC procedures on a continuing basis. For example, just prior to the thunderstorm season in Atlanta we might review procedures and phraseology associated with weather avoidance and wind shear advisories. These training days are also used to introduce new procedures or changes in old ones. We brainstorm for ideas on how to improve on the techniques currently in use within the facility and generally discuss what is going on within the agency and the facility. This focus on the review and evaluation process within the FAA is the method that a typical ATC facility uses to ensure that they are doing their part to properly exchange information and educate their controllers in all aspects of aviation.

Working for Change

What does all of this have to do with pilots and FARs? Hopefully it sets an example of how a system of checks and balances is established and how information can be exchanged, and in terms of its impact on FARs, that's easy. The specialists in the evaluation branch, working with specialists in all of the FAA's other sections, individual air traffic controllers and supervisors, dozens of employee participation groups around the country, similar groups in the private and commercial aviation sector, and individual pilots, are the people who provide the brainpower for changes in the air traffic system. That's how FARs are developed and amended.

Think about it. We are not only talking about the FAA. We are talking about the Aircraft Owners and Pilots Association, the Air Line Pilots Association, the National Business Aircraft Association, and a host of other aviation groups. When you consider all of these groups as one potential resource, you have covered almost everyone in aviation. These are the people who design the changes that are occurring in aviation. They are the people who participate in, or are responsible for, conducting in-depth evaluations of existing and proposed regulations. They study procedures to determine if new ones need to be developed or existing ones need to be redesigned, changed, deleted, or more thoroughly explained. They react to suggestions from all segments of aviation and to accidents and incidents which bring to light deficiencies in our system.

Sometimes all of these groups work well together and the result is a positive force for change in the aviation community. At other times this change is brought about by regulation, legislation, or forced adjustments caused by circumstances. But regardless of how these changes occur, and regardless of how much any particular group is opposed to a specific change, almost every aviation group supports adherence to the procedures once they become regulatory.

Your participation in any one or all of these groups gives you the ability to be a part of that change. These groups are made up of people who, like you, started out as aviation novices. They became involved for any number of different reasons, but as their involvement grew, so did their knowledge and their ability to contribute that knowledge. So the greatest benefit from this participation may be the fact that you will grow as a pilot. Remember what I said about awareness and self-teaching? If you participate in some aspect of aviation, you will become aware of what is going on in aviation, you will have to learn new things in order to continue that participation, you will have to expand your understanding of what you already know, and you will almost automatically become a better pilot as a result.

I guess I am a little bit idealistic when it comes to trying to motivate people into getting involved in aviation. I hope that I have successfully made the point that knowledge and involvement go hand-in-hand. Now, let's get back to our discussion of regulatory agencies.

Pilot Monitoring

The air traffic controller is not the only segment of aviation that comes under FAA review and quality evaluation. The Flight Standards District Office (FSDO) segment of the FAA is charged with the responsibility of monitoring the performance of the individual pilot or aviation concern. Working closely with other FAA offices and certified private-sector examiners, they monitor the performance of flight instructors by checking the knowledge and readiness of their students, and all pilots through biennial recertification programs. They perform regularly scheduled flight checks with airline and air taxi pilots, and are constantly reviewing the training programs of virtually every aviation concern in the country. This is a massive job but it is designed to ensure that every individual in the aviation community is exposed to some type of review on a regularly scheduled basis.

We frequently hear derisive comments about the agency's role in trying to maintain the high quality of aviation safety that we have come to expect. Some say that we would rather regulate than educate, but I think that this is a cheap shot. Some of the strongest advocates of requiring individual pilots to maintain their currency and their knowledge levels do not even come from the ranks of the FAA. They are the very flight instructors and examiners who have to deal with pilots on a daily basis. These private-sector aviation specialists have to teach their students as much as they possibly can in the short period of time that they have available to work with them. They must do so within a pricing structure that allows them to stay in business and still be able to say, in good conscience, that they have done their best.

These people frequently donate their time conducting safety seminars, refresher courses, participating in FAA-sponsored Operation Raincheck programs, and just sitting around talking aviation with interested people. While they might not agree with a particular regulation and will work very hard to amend that regulation to make it more effective, they, like the groups we discussed a few paragraphs ago, understand the need for regulation and wholeheartedly support efforts to regulate the quality of flight training and keep fly-by-night operators out of the aviation business.

As you can see from all of this activity, any breakdown in understanding does not come about because of a lack of effort on the part of those who are supposed to get the message out. But the fact remains that some of the message is not getting to those who should be receiving it, so let's go back to our review of some of the common misconceptions, misunderstandings, and areas where the intent of an FAR has been weakened. We will also add some lesser known bits of information that are really good to know. If you are a person who, like me, is inclined to wager a pizza on a point of disagreement, some of these little tidbits will tickle your palate and recoup the cost of this book in very short order.

STOP! PUT YOUR PENCIL DOWN

Have you figured out the brain teasers yet? If not, don't feel bad. The information regarding the single-engine airplane part of the first question and almost all of the information in the second question, with the exception of the fact that you were a radar-identified aircraft, was just useless verbiage designed to throw you off the track. I used to write questions for FAA tests and those of us who do that are fond of filling your mind with junk before we ask you to put it in gear.

If I rephrased the first question and asked 100 pilots or controllers what is the correct altitude to file for a westbound IFR flight plan below 18,000 feet, most of them could *not* tell you that *it doesn't matter which altitude you file as long as you are in controlled airspace.*

No matter how I phrased the second question, most pilots and controllers would answer by giving you a reporting list that included entering and leaving a hold, reaching or leaving an assigned altitude, and crossing the outer marker inbound on the approach. All of these are nice-to-know bits of information but *none are required by FARs.*

Surprised? If not, you are more knowledgeable about these things than the average pilot.

Let's take a look at regulatory publications in general and then analyze our questions, one at a time, to see exactly what the FARs, *Air Traffic Control Handbook*, and the *AIM* actually say about these subjects. To do that, the first thing you need to know is how to find information about a particular topic in the FARs.

FIGURING OUT THE FARS

Information regarding any specific aviation subject can usually be found in several different subparagraphs of an FAR whose title relates to the major topic under discussion. Related information in each of these FARs is then grouped together in subparts of the FAR. From there the regulations are further segregated into sections which deal with a specific subject area. Are you thoroughly confused yet? It's actually very simple, and anyone with a PhD in logic should be able to grasp the concept within a month.

Seriously though, it does make sense and we can use the information in our first question as an example of how to locate specific FARs within the regulation handbooks. In general terms, that question asks for information governing how to operate an aircraft during some phase of its flight. Essentially, we are looking for rules governing flight. This information is covered under FAR Part 91, *General Operating And Flight Rules*. Part of the title itself leads you to the major subpart, Subpart B, Flight Rules, where you can begin to narrow your search. From there a quick scan through Subpart B will lead you to the section on instrument flight rules, which directly relates to the question that I asked. The logic might seem a little obscure, and it takes some thought and a little practice to grasp it, but it really does work.

The fact that information with the same or similar subject matter is scattered throughout several different sections and paragraphs of the FARs used to cause me a little heartburn. I always felt that we should be able to group all of the information about a particular topic into consecutive paragraphs and create an easy-to-read document that didn't require you to play leapfrog while looking for information.

But, after having to read and teach this information for several years and then having to write a couple of regulatory documents myself, I have begun to understand the problem. Much of the same information or regulatory rhetoric is applicable to several different types of operations and various categories of aircraft. If one were to try and categorize information strictly by subject matter, types of aircraft, or operational activity, the resultant document would be massive and would have to repeat the same information, at length, under each subsection. So, all things considered, these books are rather well written.

(By the way, I must tell you here that, as this book goes to press, a major reorganization of Part 91 is pending final adoption by the FAA. So, while most of the FAR section numbers I quote may have changed by the time you read this, the regulations themselves and the purpose of this discussion should still be quite valid.)

Like most of you though, I would like to see an expanded index or perhaps a ''syntopicon'' added to all of these documents. A syntopicon is a separate book which lists each area of subject matter, complete with a cross reference, and the page(s) and/or paragraph number(s) where anything on a given subject is mentioned. But, considering the number of changes that occur to each of these documents every year, it would take a world-class computer, strictly dedicated to that task, to keep up with the page changes, and we would throw away enough paper every year to cover a large house. (Commercial reprints of the FARs, such as TAB/AERO's, often contain a helpful index.)

CRUISING ALTITUDES

Now, let's get back to our first question and FAR 91, Subpart B, *Instrument Flight Rules* section. If you refer to paragraph 91.121 of this section, you will find the rule governing IFR cruising altitudes and flight levels. You should note that this regulation is further broken down into subparagraph (a), which deals with controlled airspace, and subparagraph (b), which deals with uncontrolled airspace. Note that the only place where the cardinal altitude concept of even/odd MSL altitudes is mentioned is with respect to level cruising flight in uncontrolled airspace. About the only places in the U.S. where you will find large amounts of uncontrolled airspace which would allow a flight of any reasonable distance are the mountain and desert areas west of the great plains. Since, in the first question, you are operating your aircraft in the eastern part of the U.S. and in controlled airspace, the governing regulation in this question is subparagraph (a). This says that you "shall maintain the altitude or flight level assigned that aircraft by ATC."

So, can you legally *file* for 9000 on a westbound flight? Yes, you can.

Will you be able to fly your route at that altitude? Not likely.

Air traffic controllers also have regulatory handbooks which govern how they will conduct their operations, and these rules are just as binding on them as the FARs are on pilots. The primary handbook by which we define procedures (our ATC "bible") is Handbook 7110.65, which is appropriately titled, *Air Traffic Control*. Chapter 4 of this handbook deals with IFR operations. Section 5 of that chapter covers altitude assignment and verification, and paragraph 4-40 reads, "Clear aircraft altitudes according to table 4-40." That table states that aircraft operating above 3000 AGL and below flight level 290 and on a magnetic course from 0° through 179° shall be assigned an odd cardinal altitude or flight level at intervals of 2000 feet. In other words, 9000 feet rather than 8000 feet. Those aircraft on a magnetic course of 180° through 359° shall be assigned an even cardinal altitude or flight level at intervals of 2000 feet. In this case you will be given 8000 feet, not 9000 feet. So, yes, you can file for whatever altitude you want, but unless there are legitimate meteorological or aircraft operational limitations which would qualify you for an exception, the controller will normally bump you up or down to the appropriate cardinal altitude whenever your flight will operate into another facility's airspace.

Although these two regulatory paragraphs seem to be at cross purposes to one another, they really are not. The wording in FAR 91.121 subparagraph (a) still allows a pilot to file for a nonstandard altitude when he or she will be operating completely within the airspace owned by one facility. If it read like subparagraph (b), this flexibility would not exist. It also allows individual air traffic facilities to negotiate letters of agreement among themselves which permit the transfer of aircraft at nonstandard altitudes. Here again, this allows facilities to establish specific inbound and outbound routes using all altitudes without causing the pilot to be in violation of FARs while conducting this operation. So, what looks like a fouled-up mess is actually some pretty sharp minds at work.

POSITION REPORTING

Question two is a horse of a slightly different color. Here is a classic case of a good operating technique that has evolved into a perceived regulation. In this situation, almost everyone thinks that a long list of nice-to-know position reports is required to be given to ATC. This has probably come about over the years by accident as a small group of flight instructors taught the practice to a few dozen students who then became flight instructors. A successive group of instructors have expanded the teaching to the point where some of today's senior airline pilots would be willing to argue with you over the point.

But the fact remains that the only *position reports* required by ATC are those mandatory reporting reports depicted by black triangles on the enroute chart. Even these reports are only required when the flight is being conducted in an IFR, nonradar environment. There are several other pieces of information that constitute required reports, so let's take a look at exactly what the FARs say about this situation and then review the appropriate *AIM* paragraphs that seem to add fuel to the fire of disagreement.

Required communications procedures are covered in the same FAR section that we were dealing with before. FAR 91, Subpart B, *Instrument Flight Rules* section, paragraphs 91.125 (IFR radio communications) and 91.129 (operation under IFR in controlled airspace; malfunction reports).

FAR 91.125 lists three reporting requirements that include the following:

> (a) The time and altitude of passing each designated reporting point, or the reporting points specified by ATC, except that while the aircraft is under radar control, only the passing of those reporting points specifically requested by ATC need be reported;
>
> (b) Any unforecast weather conditions encountered; and
>
> (c) Any other information related to the safety of flight.

FAR 91.129 also lists required reporting items, but these are limited to malfunction reports and the information required when reporting those malfunctions:

> (a) The pilot in command of each aircraft operated in controlled airspace under IFR, shall report immediately to ATC any of the following malfunctions of equipment occurring in flight:
>
> (1) Loss of VOR, TACAN, ADF, or low-frequency navigation receiver capability.
>
> (2) Complete or partial loss of ILS receiver capability.
>
> (3) Impairment of air/ground communications capability.
>
> (b) In each report required by paragraph (a) of this section, the pilot in command shall include the—
>
> (1) Aircraft identification;
>
> (2) Equipment affected;
>
> (3) Degree to which the capability of the pilot to operate under IFR in the ATC system is impaired; and
>
> (4) Nature and extent of assistance he desires from ATC.

The above listed items plus any information requested by an individual air traffic controller are the only pilot reports required by FARs. Where then did so many pilots and controllers get the idea that there is a long list of other information required? Let's look for the answer in the *AIM*, paragraphs 341 and 342.

Paragraph 341 essentially recaps the information contained in the FAR paragraphs listed above, and it lists some recommended techniques for reporting that information. It is also a rather long section, so I will let you read it at your leisure and not reprint it here. I include this reference because of the fact that the sequencing of the information tends to show a continuation from one paragraph to the next. This arrangement, combined with the way the paragraphs are written, easily leads the reader to conclude that they are simply reading about a different category of required reports and not a separate list of highly recommended procedures. Let's review paragraph 342 in its entirety because of the fact that a large percentage of pilots and controllers consider the information in this paragraph as being regulatory, and this is the real source of the confusion.

> 342. ADDITIONAL REPORTS
>
> (a) The following reports should be made to ATC or FSS facilities without a specific ATC request:
>
> (1) *At all times*:
>
> (a) When vacating any previously assigned altitude or flight level for a newly assigned altitude or flight level.
>
> (b) When an altitude change will be made if operating on a clearance specifying VFR ON TOP.
>
> (c) When *unable* to climb/descend at a rate of at least 500 feet per minute.

(d) When approach has been missed. (Request clearance for specific action; i.e., to alternative airport, another approach, etc.)
(e) Change in the average true airspeed (at cruising altitude) when it varies by 5 percent or 10 knots (whichever is greater) from that filed in the flight plan.
(f) The time and altitude or flight level upon reaching a holding fix or point to which cleared.
(g) When leaving any assigned holding fix or point.

NOTE-The reports in subparagraphs (f) and (g) may be omitted by pilots of aircraft involved in instrument training at military terminal area facilities when radar service is being provided.

(h) Any loss, in controlled airspace, of VOR, TACAN, ADF, low-frequency navigation receiver capability, complete or partial loss of ILS receiver capability or impairment of air/ground communications capability.
(i) Any information relating to the safety of flight.

(2) *When not in radar contact*:
(a) When leaving final approach fix inbound on final approach (nonprecision approach) or when leaving the outer marker or fix used in lieu of the outer marker inbound on final approach (precision approach).
(b) A corrected estimate at anytime it becomes apparent that an estimate as previously submitted is in error in excess of 3 minutes.

(b) Pilots encountering weather conditions which have not been forecast, or hazardous conditions which have been forecast, are expected to forward a report of such weather to ATC. (See PARA—520 - PILOT WEATHER REPORTS (PIREPs) and FAR—91.125(b) and (c).)

After having read all of that paragraph consider the following. Only the word *should* in subparagraph 342(a) gives you any clue to the fact that most of this information is only *recommended*. In the *AIM*, unlike in the preceding sentence here, the word has been given no special treatment, bolding, or italics for emphasis, and it is very easily read into the context of the sentence.

Contrast that with the italicized *At all times* in subparagraph 342(a)(1) and the inclusion of subparagraphs 342(a)(1)(h) and (i), along with all of 342(a)(2) and 342(b). These subparagraphs are restatements of FAR paragraphs and are, in fact, regulatory in nature. I think you can see how easily one can draw the conclusion that the whole paragraph belongs on the "required" side of the fence.

I don't want you to get the idea that I think these reports should be discontinued, because I do not. In fact, since so many pilots already believe that all the items listed in paragraph 342 are currently required, I think that this paragraph should be lifted into FAR 91.125. The only change I would recommend would be to change *should* to *shall*.

Over the course of my career, I have talked with several experienced professional pilots who became a little testy over my inclusion of a requirement that they report one of the items listed in paragraph 342 when I issue them a clearance. They acted as though I was insinuating that they did not know how to do their job. What was obvious to me was that they were one of the people who considered these reports as being required by FARs. I have considered inviting them for a discussion of FARs over a pizza (on them, of course) but unfortunately, and probably better for my svelte figure, there is little time for such fun in this fast-paced aviation society.

The real point to be made by this example is that you should know the difference between what is required and what is a good operating practice. In this situation, most pilots think they have to do what is the right and safe thing to do. Perhaps we should not rock this particular boat since we don't want pilots to believe that they should not do what is not required.

SPEED REGS

Another regulatory jungle where we frequently see pilots using the wrong information at the wrong time is in the area of speed control and speed requirements. Here again, it is not entirely the fault of the poor, information-saturated mind behind the control yoke. If you look at the subject from the point of view of pilots, who frequently fly into several different types of airspace, there seem to be so many speed rules and so many exceptions that you need a score card to tell the plays and the players.

Actually, in the realm of the average pilot, there are only two major FARs which speak specifically to aircraft speed. (I don't mean to discount those FARs that govern air carrier and air taxi operations. Nor do I mean to exclude the operations handbooks for each individually certificated air commerce operation, because they may, and frequently do, have specific restrictions regarding speeds. Most of these documents also have requirements for speeds at which individual aircraft may operate in a given environment, but they are company policy guidelines and not FARs. Additionally, every owners manual for every aircraft has an entire section on aircraft speeds required or suggested in a given environment.) These can be found in FAR 91. One of these, FAR 91.55—*Civil aircraft sonic boom*, only talks about the requirement to operate below the speed of sound. Your typical Beech pilot doesn't need to worry about this, and even "Top Gun" would have trouble reaching those speeds, although he might give it a try. The real problem though, seems to come from trying to determine what speeds are "authorized" by ATC, when they are authorized, and when they are not.

FAR 91.70—*Aircraft Speed* reads as follows:

> (a) Unless otherwise authorized by the Administrator, no person may operate an aircraft below 10,000 feet MSL at an indicated airspeed of more than 250 knots (288 m.p.h.).
>
> (b) Unless otherwise authorized or required by ATC, no person may operate an aircraft within an airport traffic area at an indicated airspeed of more than—
>
> (1) In the case of a reciprocating engine aircraft, 156 knots (180 m.p.h.); or
>
> (2) In the case of a turbine-powered aircraft, 200 knots (230 m.p.h.).
>
> Paragraph (b) of this section does not apply to any operations within a Terminal Control Area. Such operations shall comply with paragraph (a) of this section.
>
> (c) No person may operate an aircraft in the airspace underlying a terminal control area, or in a VFR corridor designated through a terminal control area, at an indicated airspeed of more than 200 knots (230 m.p.h.).
>
> However, if the minimum safe airspeed for any particular operation is greater than the maximum speed prescribed in this section, the aircraft may be operated at that minimum speed.

There are two different types of authorized deviations from the speed requirements listed in this section. The first, "authorized by the Administrator," usually refers to a specific, written exemption for a class or category of aircraft or for a particular organization. Certain types of military operations or operations by a foreign certificated airline might be examples of this type of exception. However, when a controller says "keep your speed up," you can be reasonably confident that this is not an authorization by the Administrator to exceed the speed requirements outlined in paragraph (a) of FAR 91.70.

The biggest area of confusion related to aircraft speed assignments seems to come from the wording and the sentence structure found in paragraph (b) of this section. It is here where we find the statement, "Unless otherwise authorized or required by ATC," which is the other type of authorized speed deviation. The last sentence of this part of the regulation which, for the sake of simplicity, I am going to refer to as the *note* in paragraph (b) (even though it is not identified as a note), connects the speed requirements in an airport traffic area (ATA) to those listed in paragraph (a) when the aircraft is operating in a TCA. The wording of this paragraph allows an aircraft to operate at speeds of up to 250 knots within an ATA that is inside a TCA. Unfortunately, some pilots have taken the wording in this paragraph to mean that they must not exceed 250 knots while they are operating anywhere in a TCA. Others pilots believe that ATC can authorize a speed of greater than 250 knots at any point in a TCA. Let's look at the first segment of this paragraph and then the relationship between it and the exceptions.

By the time that most aircraft reach an ATA, controllers are usually trying to get them to slow down so that they will fit into the traffic sequence. As a result, it is a relatively rare occurrence for an air traffic controller to authorize or ask a pilot to use an excessive speed in this environment. Additionally, the vectoring techniques used by most ATC facilities to establish a sequence will keep an aircraft outside or above the ATA until they are on base or final. By this time the pilot of the aircraft has slowed on his own so that gear and flaps can be operated. Consequently, the aircraft is well below the maximum speeds described in subparagraphs (b)(1) or (b)(2).

Even the most experienced pilots misunderstand these rules, but they aren't too complicated (SEE TABLE AND DIAGRAM)

Aircraft Speed Limits*

FLIGHT SITUATION (FAR 91.70)	MAX. KIAS RECIP	MAX. KIAS TURBINE POWERED
**Within an airport traffic area (Exception TCA)	156	200
***Below 10,000' MSL	250	250
***Within a TCA below 10,000' MSL	250	250
Beneath lateral TCA limits outside TCA airspace	200	200

HOLDING (AIM)	MAX. KIAS PROPELLER	MAX. KIAS TURBOJET
Up to and including 6000'	175	200
Above 6000' to and including 14,000'	175	210
Above 14,000'	175	230

*All speed limits apply full time to ciival aircraft operated in the U.S. unless a higher minimum safe speed is required by the aircraft being operated.

**Applies unless otherwise authorized or required by ATC.

***Applies unless otherwise authorized by the administrator.

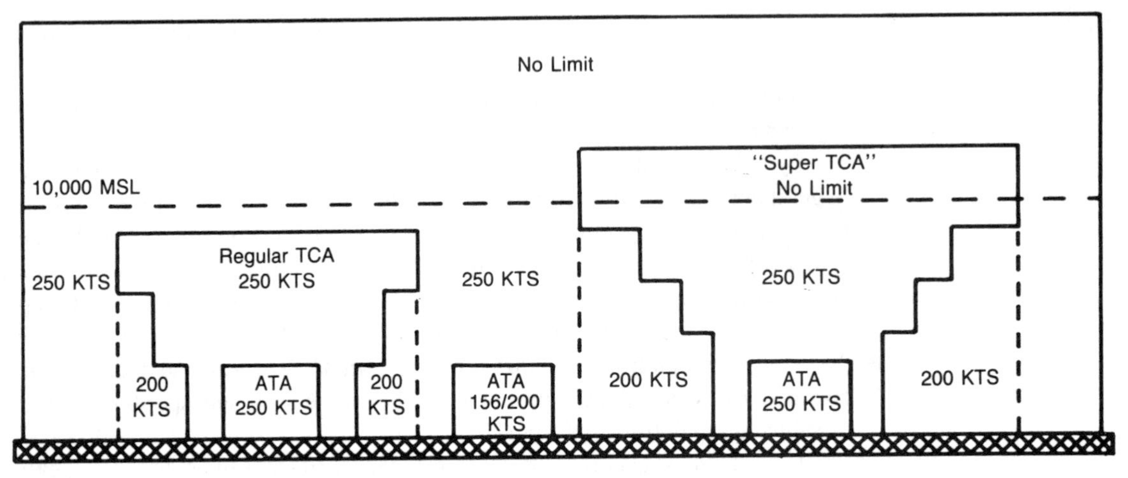

Aircraft speed limits.

- If you are above 10,000 feet, even if you are within a TCA, speed limits do not apply.
- If you are within a TCA *and* below 10,000 feet, don't go any faster than 250 knots unless the FAA's *chief* cook and bottle washer says that you can. This speed limit also applies when you are within an ATA that lies within the TCA's inner circle. It also applies if you are below 10,000 feet outside of a TCA.
- If you are in an ATA that is not within a TCA, don't go faster than 156 or 200 knots, as appropriate, unless some controller says that you can.
- If you are within the lateral limits of a TCA, but below the TCA airspace, don't go faster than 200 knots.

I was recently working a control position and vectoring an aircraft into the Atlanta airport when I instructed an aircraft to descend from an altitude of 14,000 to 11,000 (this would put the aircraft into the Atlanta TCA, while has a ceiling of 12,500 feet MSL) and told the pilot to keep his speed up. His reply was typical of the primary misunderstanding that exists among pilots with respect to speeds and TCAs. He acknowledged the descent and then said, "I understand that you are authorizing me to exceed 250 knots in the TCA." This pilot obviously took the reading of FARs very seriously or he would not have known enough or been concerned enough to ask that question. This response also made it appear to me that he was one of those pilots who believed that the *note* in paragraph (b) extended the 250-knot requirement mentioned in (a) to include the entire TCA—even the portion above 10,000 feet MSL.

Without getting into a long-winded debate with him over the interpretations of FARs, there was really no way for me to convince him that his understanding of the paragraph was incorrect. My reply to him was that he could keep his speed "at whatever he was comfortable with." He seemed happy with that answer, and the aircraft kept its ears laid back. I don't know whether this pilot believed that he was given an "authorization" to exceed the 250-knot rule or whether he was just trying to lay the monkey on my back. But, in either case, he was not totally familiar with the regulation.

If you read this FAR very carefully you can see that the *note* in paragraph (b) only says that the speeds applicable in an ATA do not apply to operations within a TCA. It makes no mention of adding any other speed requirements to operations within a TCA other than to mention TCAs in the same context where you should comply with paragraph (a) while operating below 10,000 feet. My understanding of this rule is that you can operate, in a TCA, at a speed in excess of 250 knots if you are not below 10,000 feet, and that you may exceed the 156 or 200-knot requirements if you are in an ATA which is coincidental with a TCA.

In the Atlanta area, the first time we slow an aircraft is to establish or maintain it in a traffic sequence or prepare the aircraft for descent below 10,000 feet. Once we begin to slow the aircraft for arrival sequencing, it is highly unlikely that any controller is going to ask you to go 250 knots that close to the airport. Remember, at this point you are near an airport whose traffic demands required a TCA in the first place. When we slow you down, you are usually in transition to a landing configuration, even if that transition occurs 40 miles from the airport.

All of the interpretations that I have been able to obtain from FAA officials say that, yes, you can go faster than 250 knots in a TCA as long as you are not below 10,000 feet. Having said that, let me qualify this statement by reminding you again that FARs and the Air Traffic Control Handbook are very dynamic publications which are subject to change based on what is currently happening in today's society. At press time for this book, several major changes that affect operations within a TCA were being implemented based on the 1988 findings of the Aviation Safety Commission.

I told you when we began our discussion of speed requirements that the rules change based on the circumstances and that these changes tend to sneak up on you if you are not paying attention. Let me end this part of our FAR review by showing you a subtle example of this tendency.

Paragraph (c) of FAR 91.70 adds a little requirement, mentioned earlier, that you operate at a speed of no greater than 200 knots when you are operating under (but not within) a TCA. With this rule in mind, let's construct a scenario where a controller is vectoring you in the TCA at an assigned speed of 210 knots. In most circumstances, this type of vectoring is associated with traffic pattern sequencing for arrival at the primary airport.

Most TCAs have a layer which begins at 2500 to 3000 feet AGL and extends 15 or 20 miles from the primary airport to accommodate this type of vectoring. The airspace described by this segment will normally keep aircraft within the

TCA while they are being vectored for the arrival sequence. But occasionally, because of traffic or weather in the arrival pattern, a controller might have to vector you outside of this layer and you would find yourself under the next layer. When this happens, the controller is supposed to advise you that you are leaving the TCA. The only thing that has changed in this situation is the fact that you are now *under* the TCA. But, according to paragraph (c), you should reduce your airspeed to 200 knots. This is exactly 10 knots and 5 percent of your speed so, according to Paragraph 342(1)(e) of the *AIM*, you *should* report this change to the controller.

TCA—TERRIBLY COMPLEX AIRSPACE?

Airspace segments, such as TCAs, ATAs, federal airways, etc., are also defined in FARs, and this chapter would seem like the logical place for a discussion on this subject. But, in most cases, I think we would lose sight of the difference between the regulations defining the airspace and the regulations for operating within the airspace if we begin to delve too deeply into a discussion of this general topic right now. The next chapter will cover the aspects of what constitutes *controlled airspace*, and I think that we could make the most sense out of this by combining a discussion of rules concerned with a given segment of airspace with the definition of that airspace. I would, however, like to make one exception to this concept and continue to cover the regulations governing operations within a TCA in this chapter.

TCAs are, according to some pilots, among the most complex and most regulated (some say overregulated) segments of airspace in the country. True or not, these upside-down-layer-cake-shaped massive volumes of controlled airspace are located around some of the busiest airports in the world. There are several different regulations that cover TCA requirements, and a casual reading of these various FARs does seem to have the reader bouncing back and forth between several sections in the regulation handbooks. But let's take a look at all of this paperwork, and when we are done, I think you will find that it's not quite as complex as you might think.

TCAs—Defined

The first place that TCAs are mentioned is in FAR 71, *Designation of Federal Airways, Area Low Routes, Controlled Airspace, and Reporting Points*. This FAR is simply the regulation which defines the various types of airspace within which we operate our aircraft. You might refer to FAR 71, and its companion regulation, FAR 73 *Special Use Airspace*, as ones which lay out the dimensions of the playing field.

FAR 71.12 *Terminal control areas*, defines TCAs as segments of "controlled airspace extending upward from the surface or higher to specified altitudes, within which all aircraft are subject to operating rules and pilot and equipment requirements specified in FAR 91 . . .''. It also refers you to FAR 71, Subpart K, which lists the various TCAs and details the airspace they comprise. If you are like most of us, the only copy of the FARs that you carry is the one in your approach plate handbook or a small, paperback annual *AIM/FAR* reprint. The FAR sections of these books do not normally carry a complete listing of the information that is contained in Subpart K. Nor does the FAA's own three-ring edition. TCA specifications are published in the *Federal Register*, but are not reprinted elsewhere.

So, how are you supposed to know all of this information? The answer is almost as simple as the question. The *AIM* and FAR 91.90 carry a listing of TCAs at which the pilot must hold at least a private pilot certificate to operate to/from the primary airport. As for all of the information in Subpart K, this is depicted on sectional and terminal area charts. These charts also carry, either in the margin or on a legend panel, a listing and explanation of most of the symbols used on those charts.

A lot of pilots that we talk to on the frequency act like this information is Greek to them, so remember the following. You are supposed to study this information either in class or at home. If you wait until you are at the controls of a high-performance aircraft to review this data, you have a very foolish person for a pilot.

In recapping FAR 71 we find that this section on TCAs is simply a definition of airspace referenced in other FAR parts. FAR 91 is where the real requirements for operating in a TCA are located.

TCAs—The Rules

Just about all of the information that you would ever want to know about TCAs is located in FAR 91.90. This section is broken down into four paragraphs. Here again, the section is quite long and reprinting it in its entirety would serve no purpose. I will simply provide a synopsis and touch on the important information in each paragraph.

Paragraph (a), concerning operating rules, contains a subparagraph which advises pilots that they may not operate an aircraft within a TCA unless they have received ''an appropriate authorization from ATC prior to the operation of that aircraft in that area.'' What this means is that a controller has to say ''cleared into,'' ''cleared out of,'' or ''cleared through'' the TCA *before* you place your aircraft in that airspace.

The wording of this subparagraph seems reasonably simple to me, but some pilots like to stretch the limits of credibility just a little bit. We frequently see a pilot allow his/her aircraft to penetrate the TCA without such a clearance and then hear the argument that they were trying to talk to the controller but he or she was too busy to answer. Some of the other arguments that we hear about why an aircraft gets into the TCA are impressive, and some are creative enough to suggest that the pilot should be a lawyer or a writer of high drama and fiction. But even if your arguments would impress an administrative law judge, frequent unauthorized flights into congested, heavily traveled TCA airspace have the potential of bending more than just the rules. I seriously doubt that Saint Peter would be overly impressed with your ability to argue before the bar.

The next subparagraph contains a provision requiring pilots of large, turbine engine-powered aircraft, who are landing at or departing from a primary airport within the TCA, to operate at or above the floor of the TCA while they are within the lateral limits of the TCA unless otherwise authorized by ATC.

There are two points to be made in connection with this subparagraph. First, the classification *large, turbine engine-powered aircraft* is not limited to jet airplanes. *Large* aircraft are those whose maximum gross takeoff weight exceeds 12,500 pounds, and *turbine engine-powered* aircraft includes turboprops. Second, while there may be many airports that are within the lateral limits of a TCA, and several within the TCA's inner "ring," this section speaks only to those aircraft operating to or from a *primary* airport, an airport around which a set of TCA rings is centered. (A new "Super-TCA" may have several primary airports.) Technically, this rule does not exist for operations into or out of other airports, and pilots of this type aircraft are not bound by this rule when they operate into those airports. Regardless of whether or not it is a very intelligent thing to do, pilots of large, fast, turbine-powered aircraft can legally scoot along underneath the lowest sections of a TCA, mixing it up with the puddle jumpers of the world. The other paragraphs in FAR 91.90 outline the pilot and equipment requirements for operating in TCAs and these sometimes cause confusion among novice pilots. Paragraph (b), *Pilot Requirements*, states:

> (1) No person may take off or land a civil aircraft at an airport *within* a terminal control area or operate a civil aircraft *within* a terminal control area unless:
>
> (i) The pilot-in-command holds at least a private pilot certificate; or,
>
> (ii) The aircraft is operated by a student pilot who has met the requirements of § 61.95.
>
> (2) Notwithstanding the provisions of (b)(1)(ii) of this section, at the following TCA primary airports, no person may take off or land a civil aircraft unless the pilot-in-command holds at least a private pilot certificate . . . [the list of these airports, formerly classified as Group I TCA airports, currently includes ATL, BOS, ORD, DFW, LAX, MIA, EWR, JFK, LGA, SFO, DCA, and ADW].

I have emphasized the word *within* because it is the key word in this regulation. Only those airports that are located within the area where the TCA extends upward from the surface (i.e., the inner ring) can be considered as being *within* the TCA.

There are, of course, hundreds of airports that lie under the umbrellas formed by TCAs (but not within the inner ring), and these airports are not affected by this paragraph. Although student pilots may fly solo around or under a TCA without any special TCA training, I would caution that this type of congested airspace is not the type of proving grounds to test your mettle or prove a point. Obtain some high-quality flight instruction and conduct several flights with experienced pilots into this environment before you attempt to operate solo near such a high density traffic area. I also do not mean to suggest that you avoid this area out of a sense of fear. In this case however, caution and experience go hand-in-hand with safe flying.

In order to fly solo through a TCA, or to land or take off at an airport *within* a TCA (within the inner ring—whether or not the airport is a *primary* airport), a student pilot must receive the TCA training and logbook endorsements specified in FAR 61.95. Even with this training, solo takeoffs and landings by student pilots are prohibited at 12 of the busiest airports.

Paragraphs (c) and (d) define the three simple equipment requirements associated with operation in a TCA. First (except in the case of helicopters prior to July 1, 1989), you must have an operable VOR or TACAN. Second, you must have a two-way radio capable of communicating with ATC on the appropriate frequencies. Third, you must have an operable, Mode C equipped transponder to operate within the TCA. Effective July 1, 1989, this last requirement is extended to operations within a 30-mile radius of each TCA primary airport, from the surface up to 10,000 feet MSL.

Refer to FAR 91.24 for details of transponder requirements. FAR 91.24, when combined with FAR 91.90, simply says that you must have a functioning Mode C transponder unless ATC gives permission to operate without one or you meet certain exceptions (glider activity, balloons, etc.).

In the case of a transponder failure, we (ATC) can authorize such an operation regardless of when the transponder actually failed as long as prior approval has been given for that operation. As you would probably suspect, we tend to keep track of the aircraft that come in and out of the TCA without a transponder and will withhold such an authorization from an aircraft that tries to circumvent the requirements routinely.

In the case of a failure of a required navigational or communication component, we can authorize continued flight into, out of, or through the TCA if that failure occurred in flight. The section requiring communication on "appropriate frequencies" does not necessarily mean that you have to have a radio that will transmit and receive on every frequency that we use. Remember though, a controller is cautioned not to work aircraft in another controller's sector without proper coordination, and we tend to decline operations that force those kinds of circumstances. If you are unable to communicate with a controller on the appropriate sector frequency, or that sector is very busy and the controller does not have a usable lower-frequency-range selection to assign, you may find your VFR operation in the TCA declined for reasons of traffic workload.

What I hope you can see from these last few pages is that TCAs are not as complex as you might have feared. Operations in the vicinity of TCAs are similar in many respects to flying in any other area except that they are usually more congested. One fact that might put this into a better perspective and allay the fears of novice pilots is that, just like any other area in the United States, these areas also have flight training facilities complete with very novice (or was that nervous) pilots.

The average student who begins his/her flight training near a TCA location usually brings the same skills and abilities to their first lesson that any average student in Medianville, U.S.A. might have. Given these circumstances and standardized flight training, it is logical to assume that both students would develop into pilots with relatively equal abilities. The only advantage that the one would have over the other would be familiarity with the local environment. A pilot who just completed flight training in the Nevada mountains need not be in awe of someone who operates in and out of a busy TCA in the Atlanta or Chicago area. In fact, there are a large number of pilots with several hundred hours of flight time, all obtained in the flat lands around these TCAs, who could learn several lifesaving lessons about mountain flying from this new pilot.

But just because you are a confident and competent TCA jockey doesn't mean that you should bore holes in the Nevada mountain air (or vice versa). A few hours spent learning the local area from a knowledgeable native is invaluable. It is also frequently all that a good pilot needs to do to become as competent as the locals in the new geographical area.

ATC INSTRUCTIONS VS. PILOT AUTHORITY

In a very simplistic sense, there are two fundamental FARs which establish the framework for authority under which pilots and controllers make the ATC system work. The first, FAR 91.3, *Responsibility and authority of the pilot in command*, states that the pilot in command "is directly responsible for, and is the final authority as to, the operation of" the aircraft. It goes on to say that the pilot may deviate from any of the general or flight rules listed in FAR 91 to meet an emergency although he or she may be asked to explain those actions. The wording of this FAR is intentionally broad so as to allow the pilot the maximum flexibility in emergency situations. Yet this authority is effectively limited by the twin requirements of responsibility and possible accountability. Historically, pilots have used this rule wisely, and the ATC system has been able to react easily to those few cases where pilots deviate from standard ATC procedures.

That leads us to the second rule, FAR 91.75, *Compliance with ATC clearances and instructions*. This rule establishes the policy that, except in an emergency, no pilot may deviate from an ATC clearance once that clearance has been obtained. It further states that, "except in an emergency, no person may, in an area in which air traffic control is exercised, operate an aircraft contrary to an ATC clearance." As with the paragraph concerning pilot authority, this paragraph advises that a pilot who does deviate from ATC instructions in an emergency may be asked to explain those actions. It also states that any pilot who deviates from ATC instructions must notify ATC of that deviation as soon as possible.

These rules are designed to complement each other, give each of the participants the authority to react to an emergency, and provide everyone the confidence that all parties will do what has been agreed upon. Unfortunately, the average air traffic controller can tell you a few war stories about some excessively liberal pilot interpretations of FAR 91.3.

Surprise Again!

When an air traffic controller issues an altitude assignment or a navigation route to a pilot, we expect, and are frequently predicating separation on, that pilot's compliance with those instructions. If the flight is being conducted in IFR conditions and/or on an IFR flight plan, we fully expect that pilot to punch through a cloud or get the wingtips wet if that is what it takes to adhere to a clearance. I don't mean to sound uncaring, because I'm not. We will gladly help a pilot vector around weather or assign another altitude if that is what it takes to provide a better ride and we are able to do so. But unless you have been authorized to make a deviation or are experiencing an emergency about which you have informed ATC, you "shall" comply with ATC instructions.

It is a very frightening thing to scan your radar scope back and find an aircraft as much as 90 degrees off course from its assigned route. When we ask the pilot what is going on, we frequently get a very cavalier answer. "We're just dodging around a little buildup." The fact that this aircraft is headed into another controller's airspace at 450 knots doesn't seem to concern the pilot, and he or she is totally oblivious to the fact that they are now on a conflicting course with another aircraft at their altitude.

Another similar situation which bears mentioning in this context is a discussion I had with a pilot who was flying 500 feet above his assigned altitude. When I asked him what he was doing, he very brazenly responded that the assigned altitude had put him in the cloud tops so he had "popped up a bit to get out of the bumps."

Pilots don't seem to know or care that these are technically violations of FARs. If this type of activity were limited to just a few pilots, I think that we could solve the problem easily. Unfortunately, it seems that a large number of pilots have taken the position that their authority for the safe operation of their aircraft includes the avoidance of even the slightest potential for trouble and/or discomfort.

Several years ago I rode in the cockpit of a commercial aircraft and watched what I must assume is typical of this kind of operation. We were cruising at flight level 310, and the last ATC instruction was "fly heading 340 until receiving XXX VOR suitable for navigation, then proceed direct." The course we were on was going to take us within a mile or two of a rather large buildup, and the pilots were discussing the fact that these types of clouds were beginning to spring up all over to the right of the aircraft. The pilot of this aircraft reached over and set the heading indicator bug on his autopilot to 320°. The aircraft made a lazy turn away from the buildup, and I looked down at the VOR indication on the flight director which had begun to show a solid, centered indication.

About four or five minutes later the controller called and asked the pilot to say his heading. The pilot responded "340°". By this time we were abeam the buildup, and the VOR indication was showing, and had been for about two minutes, a full right deflection. The controller replied that "the wind must be higher than I thought at that altitude, are you receiving XXX VOR yet." The pilot confirmed that they were and advised that he could now proceed direct. I have to assume that either the pilot of this aircraft did not feel what he was doing was wrong or he did not think I was knowledgeable or smart enough to know what he was doing. Since I do not know what was going on in the pilot's mind, I gave him the benefit of the doubt and decided to be a little more cautious of the separation that I use when pilots are deviating around buildups.

Increasingly, pilots are taking unannounced, evasive action around virtually all types of weather. In some cases, they are flatly refusing to take controller-assigned headings that they think will involve them in a weather system. While we are as familiar with the cases of weather-related accidents as they are and we understand their reluctance to come within miles of a buildup, they put us in a very serious bind when they refuse to follow instructions. Ask any pilot in this situation if they are declaring an emergency and most of them will say, "of course not." Yet every one of these pilots is not going to go where the controller needs for them to go, and almost every one of them seems to want to take a different path around the weather. As the traffic volume and the weather systems build, the result can become chaos and we start having to hold aircraft out of our airspace or on the ground.

Those pilots who are on the ground start chirping about delays and tell controllers that they can take whatever headings are needed if they can just get in the air. When you launch this guy, his first transmission to the departure controller is frequently a statement that he wants a turn away from the weather. Sometimes they don't even ask, they just turn.

Controllers often find themselves between a rock and a hard place in these situations because they do not have control, yet they are still responsible for separating this pilot from other aircraft and from other controllers' airspace. We know that the pilots are only trying to do what is safest for their aircraft. Unfortunately, their insistence on doing this, without accepting the companion responsibility of exercising their emergency authority, places the controller in the position of having to play by a different set of rules.

We can and, if necessary, we will bend or stretch a few rules to accommodate an emergency situation. But if you insist that you are not exercising emergency authority, we must keep you in our airspace or coordinate the use of someone else's. We must still keep you the minimum distance and altitude away from other traffic regardless of what that requires and we must coordinate your nonstandard activities with the next controller who is going to have to work you. We must frequently do all of this without having the foggiest idea of what you are going to do next. So, if you ever wonder why a controller seems overly cautious, think back to these situations. As the old adage goes: Fool me once, shame on you. Fool me twice, shame on me.

Just about every controller has a pet peeve regarding the things that some pilots do and this discussion about pilot authority and compliance with ATC clearances highlights mine. But it also serves to point out that the application of FARs must deal with the reality of the situation in which pilots and controllers find themselves. If a controller took a hard line every time a pilot felt it necessary to avoid a cloud, we would very quickly develop an adversarial relationship between pilots and the controllers in that facility. Such a relationship would, in very short order, spill over to other facilities and other groups of pilots, and the entire system would start to break down. As a result, we (both pilots and controllers) allow a certain amount of tolerance in those cases where we feel that one of us has stretched a rule or procedure just a little bit. Additionally, pilots and controllers have developed a set of operating techniques over the years which, while not exactly according to the strict application of FARs, does work in the real air traffic control world. One example of this situation is in the airfiling of IFR flight plans.

IFR AIRFILING

We talked at some length in Chapter 2 about what information a controller actually needs to process a flight plan in our computer systems. I also mentioned at that time that the pilot is required to give ATC much more information than a controller actually needs to do that job. FAR 91.115, *ATC clearance and flight plan required*, advises that a pilot must obtain an ATC clearance and file a flight plan before he or she can operate under IFR in controlled airspace. (Does this mean that you don't need to do this if you are flying in uncontrolled airspace? Check this one out with the experts.)

There is also a regulation, FAR 91.83, *Flight plan; information required*, which tells a pilot what information has to be given to ATC when filing a VFR or IFR flight plan. The first section in this paragraph states that a long laundry list of eleven items of information must be given to ATC unless they authorize otherwise. Normally, a controller needs less than half of this information to actually process the flight plan in our system, and when pilots tell us that they want to airfile, we usually just say "go ahead with the flight plan." Since this action does not meet the, *unless otherwise authorized* release from the requirements as stated in the first paragraph of this FAR, pilots should give the controller the complete list of all flight plan elements.

Controllers who are familiar enough with the technical requirements of the FARs realize that they must ask pilots only for the "needed" information if they want a reply limited to that information. Some pilots have learned the list of items that a controller needs, and they limit their transmission to those items. Technically, this action is a violation of FARs and pilots run the risk of not having important elements of the flight plan information on file in the event of an accident. Furthermore, all pilots are required to have a weather briefing before their flight and, unless they obtain one from the controller (who rarely has the resources or the time) or get one from the FSS without filing a flight plan (which is a waste of everyone's time), there is no record of them having done this.

There are two possible solutions to this dilemma. First, in the event of an airfile with an air traffic control facility other than a flight service station, pilots could reduce the verbiage with ATC by asking the controller if they want only that information that goes in their computers. Second, and to my mind the more preferable of the two, all pilots should obtain a weather briefing and file an IFR, or for that matter VFR, flight plan with the FSS before every flight. They should, of course, remember to contact the destination FSS to cancel all VFR flight plans and any IFR flight plans that do not terminate at a controlled airport. Any ATC facility can cancel these flight plans, but VFR flight plan cancellations have to be relayed to the destination FSS. With this in mind, be sure to cancel on your own, or at least have the courtesy to ask the controller if he or she has time to do this for you, since it's extra work for the controller.

There are many other FARs that we could discuss which are, or seem to be, controversial, ambiguous, or very subtly misunderstood. But, I think that by now you should have begun to understand that you can't just gloss over FARs in your studies. You need to learn them as thoroughly as possible, and you should use every resource at your disposal to understand them.

LEARNING ATC PROCEDURES

ATC procedures are the other area of knowledge misunderstood by a large number of pilots, but pilots are not normally expected to know a great deal about them. They present some of the same types of contradictions that are found in FARs and have the added disadvantage of being information that is not readily available to pilots. Our procedures handbook is about the size of an unabridged dictionary and we incorporate dozens of changes into it each year. Very few pilots or instructors, and only a few flight schools, subscribe to this document, and as a result, very few pilots have actually read copies of the literature that is our bible of air traffic control.

Some of the air traffic procedures that we use can also be found in the *AIM*, which is a very good, general information publication that covers almost all aspects of aviation. But these specific topics, such as air traffic procedures, are scattered throughout the book and sometimes tend to get lost among all of the other information it contains. Just like the FARs, which are also referenced in the *AIM*, air traffic procedures are best learned by reading a book dedicated to that subject.

We have commented on several air traffic procedures as we have gone through the last few chapters, but I think that a few others bear mentioning. Let's go over some of these procedures and relate some stories that will illustrate how a little understanding will go a long way toward a more efficient and sometimes shorter flight on your part.

ASK AND YE SHALL RECEIVE

There are some magic words in the ATC language which, if a pilot knows (and knows how and when to use them), will get the pilot what he wants, when he wants it. These words are magic in the sense that they are the triggers that allow controllers to authorize procedures that they are not permitted to suggest or even hint as to their existence.

The logic behind establishing procedures such as these is very sound in that the act of asking for authorization to conduct a procedure tends to demonstrate a pilot's knowledge of that procedure. If you remember the discussion we had on Special VFR operations in Chapter 3, you know that this understanding may not always exist, but this is an example of the type of operation about which I am referring. Let's look at the general concept of pilot-requested operations and then at some other examples of ATC procedures that you should know.

Individual FARs often begin with the statement, *unless authorized by ATC*. When used in its proper context, this qualifier allows a pilot to deviate from certain regulations within prescribed limitations or to do something that would otherwise not be approved.

This same type of qualifier exists in ATC procedures except that the wording *if specifically requested by the pilot* places the requirement for initiating the request on the pilot. This statement is found frequently in our procedures handbook, and in virtually every case it refers to a situation where there is a reduction in the separation, distances, time, or procedural minimums that we would normally have to apply. A request for these reductions indicates to the controller that the pilot knows what will transpire in this instance, is prepared to conduct the operation, and is willing to accept the responsibility for that action and the reduced minimums or separation that go with it. Let me give you an example.

Intercepting Final

Old Sam Cessna is inbound to the home airport (elevation 1000 feet) on an IFR flight plan, and he is in a very big hurry. The airport is VFR with a reported 2000-foot ceiling (i.e., the ceiling layer is at 3000 feet MSL), the surface visibility is only three miles, and the inflight visibility is less than that. The controller advises Sam that other pilots have been unable to see the airport before they would have penetrated the airport traffic area and, as a result, the controllers have been conducting ILS approaches. Sam is smart enough to know that, because of the local traffic, he doesn't want to come charging into the traffic pattern with this kind of visibility, so he does not cancel his IFR and he tells the controller, "We'll do the ILS, but we don't want any delays."

Our ATC handbook tells us that, unless certain conditions exist, we must vector an aircraft to intercept the final approach course at least 2 miles outside of the *approach gate*. The approach gate is a point that is 1 mile outside of the outer marker or final approach fix and must be at least 5 miles from the landing threshold. In the case of this ILS approach, we will assume that the outer marker is 5 miles from the threshold, the approach gate is 6 miles from the threshold, and the turn-on point is at least 8 miles from the threshold.

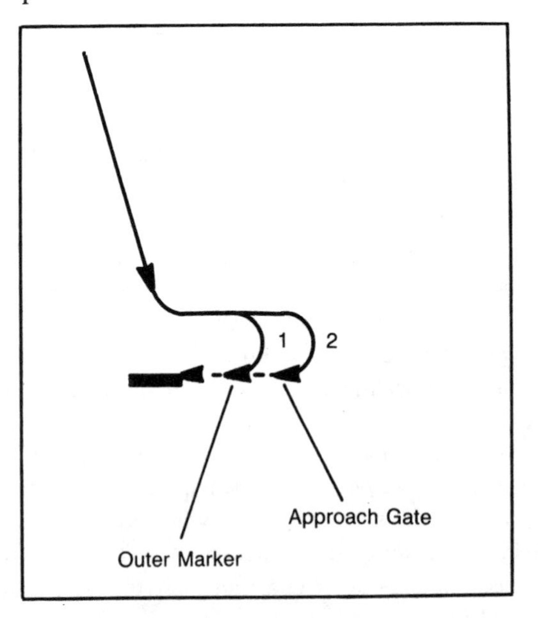

Experienced pilots can save time and fuel by requesting a reduced intercept interval when being vectored to final approach. Route 1 would have been Sam's route if he had asked the controller for a turn over the outer marker. Instead he was vectored to final two miles outside the approach gate, which is one mile outside of the outer marker (Route 2).

The intercept distance can be shortened (those ''certain conditions'' I mentioned in the preceding paragraph) if the right combination of events, requests, or environmental conditions exist. The first of the potential exceptions states that the controller can vector to intercept the final approach course at the approach gate if the weather is at least 500 feet above the Minimum Vectoring Altitude (MVA) and the visibility is at least 3 miles. In this case the visibility is 3 miles but the ceiling layer is only at 3000 feet MSL, and the MVA in this location is 2700 feet MSL. Since we require the ceiling layer to be at 3200 feet MSL or above to meet the criteria in this exception, a controller cannot shorten the intercept point based on good weather conditions.

The second condition states that we can vector to intercept the approach course as close as the final approach fix if such an intercept is specifically requested by the pilot. Such a request would take a form similar to the following; ''Cessna 5TA will take a turn-on at the marker.''

That's not even close to the phrase that Sam used, so the controller is forced to vector 2 miles outside of the approach gate for the intercept.

If the controller has enough of an indication from the pilot that he is asking for a reduced intercept interval, the controller will normally try to solicit the ''specifically requested'' statement with a tactfully phrased question. Such a question might go something like: ''Cessna 5TA, say again how you would like us to conduct the approach?'' Unfortunately, this question often brings about the same or a similarly useless answer, or the pilot may make some caustic reply and infer that we are not paying attention.

We are required to provide a pilot with an intercept angle to the final approach course of no greater than 30° so that he can establish on final with at least 1 flying mile of level flight before he is required to descend the aircraft. Given this requirement, the actual flight path that Sam will fly will be a base leg at about 9 to 9.5 miles from the threshold. He will fly an additional 6 miles or more beyond where he could have turned base had he known enough about ATC procedures. So, as the old saying goes, you get what you pay for, and Sam just paid for an 8-mile intercept when he could have had a 5-mile intercept.

The Contact Approach

What happened to Sam is actually somewhat mild when you consider some of the other things that pilots have had to do because they didn't know enough about ATC procedures. I once worked an aircraft that was trying to get into a remote airport where the only published approach was a nondirectional beacon (NDB) approach. The aircraft was not equipped with ADF equipment, but the pilot wanted to fly over the airport in hopes that he could conduct a visual approach and land. The pilot reported that the weather at the airport was marginal with a broken layer of clouds just below the MVA, but advised that the visibility in the clear areas was excellent.

The pilot told me that he had a very important meeting to attend, so I vectored him over the airport several times, but each time he was at a point where he should see the airport, he was in the clouds and could not. He kept telling me that he could see the ground and that he knew where the airport was located, and asked me if I could give him a lower altitude because he was certain the he could get to the airport if he could just get below the clouds. I advised him that he was currently at my minimum vectoring altitude and I could not authorize a lower altitude while he was on an IFR flight plan. The pilot eventually gave up and flew to another airport to land, and I assume, rented a car to drive to his meeting.

This pilot obviously did not know about the one type of approach that was available to him in this situation, the *contact approach*. This procedure can be an alternative to an instrument approach on an IFR flight plan when the reported ground visibility at the destination airport is at least 1 statute mile, and the pilot has *requested* a contact approach. When conducting this type of approach, the pilot is responsible for maintaining 1 mile flight visibility, staying clear of clouds, maintaining legal terrain/obstruction clearance, and navigating whatever route to the airport is needed to maintain that clearance. In this case, the pilot indicated that he was able to do all of the required items but he did not know enough about the procedures to ask for a contact approach.

This kind of situation is just as frustrating for the controllers because they know that there is something that they can do to help if the pilot would just say the magic words. In cases like the one mentioned above, we are reasonably confident that the pilot has the skills to conduct the approach and we find ourselves almost wishing the words into the pilot's mouth. We also share the pilots' disappointment when they have to accept an expensive alternative. But our procedures are very specific about how we have to handle this situation and the wording in our procedures handbook, which states, "It is not in any way intended that controllers will initiate or suggest a contact approach to a pilot," leaves no doubt as to how we are supposed to do our job.

I have often been tempted to wait until just before I switched them to the enroute frequency at their alternate airport and then tell pilots like this to study the procedures for a contact approach after they get on the ground. But I usually start wondering about what my liability would be in this situation and I just give the pilot my operating initials and invite him to drop by the tower, look me up, and remind me that he was the pilot who had tried to get into the small airport. This way I could use my authority and credentials as an instrument ground instructor to actually teach this pilot about contact approaches.

Intersection Departures—Hurry Up and Wait

If you think that a situation where pilots don't know about the procedure can be frustrating to a controller, consider the opposite side of the coin. There are situations where a pilot could know part of the requirement but not enough to allow a controller to authorize the operation after it has already started. Often they know just enough to put themselves in a position where the result is frustrating. See if this scenario sounds familiar.

One of our favorite pilots, Ms. Barbara Beech, has made a request to a ground controller, which from their standpoint could easily be approved. She has requested an intersection departure. If B.B. were the only aircraft involved, there would be no need for further action and she would be able to go merrily on her way. way. Unfortunately, the fact that she is not the only aircraft taxiing to that runway alters the circumstances to the point where she must do one more thing. She must either initiate a request for a reduction in separation or wait until that reduction is no longer needed before the intersection departure can be approved. (A request to waive the wake turbulence requirement is our authorization to conduct this type of operation.) If B.B. does not realize this, her takeoff clearance will be delayed. Let me give you a more detailed example.

I have worked at several airports where the general aviation or cargo ramp connects to a long, main runway at an intersection which could easily be used as a departure point with several thousand feet of runway remaining (SEE DRAWING). Because of the fact that they will be faced with a long taxi route to the approach end of the runway, pilots frequently request an intersection departure at the point where the ramp intersects the runway. Controllers can, of course, approve this request and might even suggest it if there are no other mitigating factors. As long as this is the only aircraft involved or the other involved aircraft do not require in-trail wake turbulence separation, the controller can approve the

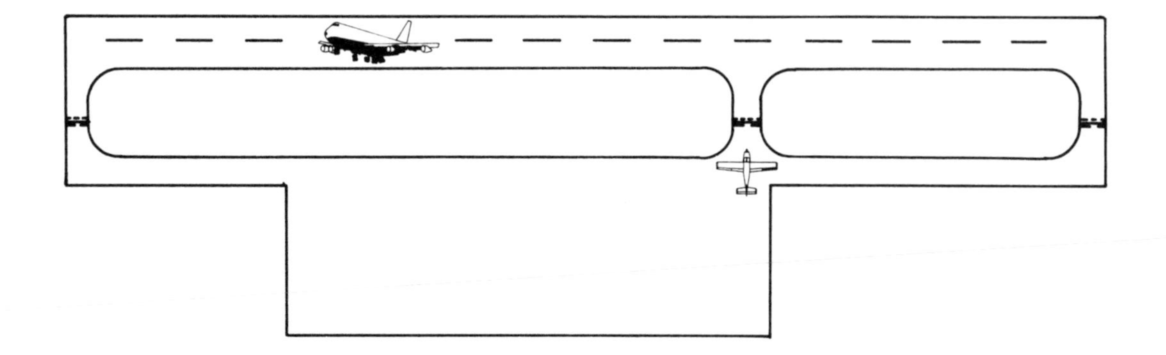

Unless you waive it, expect an automatic wake turbulence hold when taking off behind a departing large aircraft.

intersection departure without delaying the aircraft at the intersection. Now let's introduce the second factor into our equation.

If a large aircraft (a classification of aircraft weighing more than 12,500 pounds maximum certificated takeoff weight, up to 300,000 pounds) has departed from the end of the runway, we are required to advise the pilot at the intersection that there will be a 3-minute delay for wake turbulence before the pilot can depart. The pilot has the authority to waive this requirement and he or she must be the one who initiates this waiver request. Because of the fact that we are not permitted by procedures, plus the fact that such an operation is potentially dangerous, the controller cannot and will not suggest such a waiver to the pilot. (Personally I think that anyone who takes off from an intersection behind a large aircraft, even with the 3 minutes, is operating a few bricks short of a full load. The savings in fuel and time won't even pay for the insurance policy that you should carry.) In any event, if the pilot does not request a waiver, and (this is the important part) there are no other aircraft involved, the controller will simply advise the pilot about the wake turbulence delay (giving the pilot the opportunity to initiate a request for a waiver), wait the required three minutes, and then clear the aircraft for takeoff.

Now let's assume that there is a whole line of large aircraft at the approach end, waiting their turn for departure, when the pilot initiates a request for an intersection takeoff.

If the pilot fails to request a waiver, the controller is faced with three bad choices:

- The controller can delay all of the aircraft in the departure queue for three minutes so that the aircraft at the intersection can be next for departure. This is one of the best procedures that we can use if we want to irritate the maximum number of pilots at one time. (I really do find it hard to believe that a pilot would be selfish enough to expect this kind of preferential treatment.)

- The controller can continue to run departures from the runway end, which effectively delays the pilot at the intersection by three minutes for each departure that is run. If we want to enrage just one pilot, this is one of the most effective ways to do so.

- The controller can advise the pilot, who has already been told that an intersection departure could be approved, that he or she will now have to taxi to the end and get in line with everyone else. We use this procedure when we want to start pilots muttering to themselves.

Seriously though, these pilots have placed themselves in a situation where the controller, who is quite willing to expedite their operation, is now forced into an situation that will delay at least one pilot or rescind an approval that had been previously granted. Most of us have gotten to the point that we will tell a pilot who requests an intersection departure in this situation that an intersection departure *could* be approved but that there would be a 3-minute delay because of wake turbulence. If the pilot does not offer the waiver at that point, we will treat the situation as though the pilot had never asked and we will issue taxi instructions to the end.

Unfortunately, if the pilot wants to take issue with our decision to taxi their aircraft to the end, we cannot even mention the reason why we have chosen this course because it could possibly be construed as a suggestion to the pilot that they ask for a waiver. Some pilots get very testy about this because they think the controller is giving priority to the large aircraft. The controller, on the other hand, can only become frustrated about the situation and wonder if these pilots got their basic flight training from the instructions on a cereal box.

BUT DON'T ASK FOR THE IMPOSSIBLE

Occasionally a pilot will ask us for permission to do something that we are unable to authorize. Some will even go so far as to rephrase the question and ask several times while trying to get us to say something that can be construed as an approval. Most pilots probably don't realize that they are asking for something that cannot be granted, and they take the controller's disapproval as either a misunderstanding on their part or an arbitrary decision made without cause. Controllers have to be very careful what they say in these circumstances because, if something goes wrong, you can bet that this pilot's lawyer will be poring over the transcript of the incident, trying to find a way out for his/her client. (It is highly probable that our lawyers will be studying the same transcript to be sure that we didn't do something wrong.) Let me give you an example of what I am talking about.

Just about every airport has locations on it that are described as nonmovement areas. General aviation or cargo ramps, terminals, vehicle access roads, maintenance areas, and fuel storage areas are examples. These areas are portions of the airport on which the ATC facility does not and will not exercise air traffic control or issue control instructions. When you move around in this area, you are essentially operating at your own risk.

I had a pilot call me while he was in one of these areas and, as best I can recall, the following exchange took place (I have changed the call signs and location):

Pilot: *Ground, this is N5TA on the cargo ramp. We would like to taxi around over here to test out some new brakes we just had installed.*

Controller: N5TA, ground control. Be advised that you are on a nonmovement area and we are unable to authorize operations in that area.

Pilot: *Ground, we just want to reposition the aircraft on the ramp.*

Controller: N5TA, I say again that you are on a nonmovement area, and I am unable to issue an authorization for the operation that you have requested.

Pilot: *Is it OK if we just move to another gate on the ramp? We can test our brakes as we taxi.*

Controller: Sir, I am unable to approve your operation or decline to approve your operation. You are on a nonmovement area and I cannot issue control instructions for operations on that area.

Pilot: *Roger, we are just going to taxi around on the ramp a bit, and we'll call you when ready to taxi.*

I could have simply replied "Roger" to this last transmission, but I had a gut feeling that I might have to defend the use of that word in the event something went wrong. Remember, this aircraft had just had its brakes repaired. It had also landed with a hot brake earlier in the day and had required assistance at that point. Because of these circumstances, I made the following transmission:

Controller: I have received and understood your last transmission.

There were several other pilots on the frequency when this happened and, for the next couple of minutes, every time I switched one of them to the tower frequency, they responded with, "I have received and understood your last transmission."

I found this somewhat amusing even though it was at my expense. But my point in using this verbiage was that I did not want to imply or even leave open the possibility that I was granting any type of approval for what this pilot was asking.

SAYING IT RIGHT

Controllers often have their hands tied when they are talking to pilots because there are certain situations where only the exact phraseology will suffice. There are specific procedures established for almost every situation that you can imagine and required phraseology designed for use with that procedure. We are also cautioned that if a situation occurs for which no phraseology or procedures exist, we should exercise our best judgment and use whatever phraseology exactly describes what we mean. If we use the wrong words at the wrong time we might be in for a few surprises, and we will probably not get the results that we want.

Headings

When a controller wants to issue a heading to a pilot, he first must ask himself if he knows the heading that the aircraft is currently flying. If the answer is no, he is supposed to use the phraseology, "Fly heading xxx degrees." If he knows the heading (this may be because he has previously assigned a heading to that aircraft or the pilot has told him the heading being flown), he can say, "Turn left (right) heading xxx degrees." The reason for this becomes obvious when controllers are issuing turns of only a few degrees.

Let's suggest that a controller needed to put an aircraft on a track of 320° to establish a sequence. If the controller looked at the radar scope and observed the aircraft on a track of 300°, the logical instruction would be, "Turn right heading 320." But suppose there was a significant wind out of the northeast and the pilot was already heading 330° to compensate for the wind drift. A controller's instruction to "Turn right heading 320" could, and *has*, resulted in a 350-degree turn to the right. This is one of the reasons why we say, "Fly heading."

Braking Action

The situation where we try to get the pilots to use the right words can sometimes be just as comical. We are required to report runway braking action to pilots whenever circumstances exist that may cause a reduction in the normal conditions. Our handbook advises that we are to use only four words, or combinations of those words, to describe the conditions that exist. Those words, *good, fair, poor* and *nil*, were chosen because they adequately describe the conditions that they represent and, when used in combinations, such as fair-to-poor, they illustrate the range of conditions that may be found.

Unfortunately, pilots often use words that are not among those listed, and they frequently choose ones that are very ambiguous. Here are a few of those selections:

Slick—Does that mean icy?

Just great—Sarcasm or a condition better than good?

Not bad—If it's not bad, it must be good, right?

It's all right—Talk about lukewarm answers!

I think by now you have gotten the point. Each one of these answers requires that the controller go back to the pilot, list the appropriate answers, and ask the pilot to classify the braking action according to one of the usable answers.

Sometimes even this doesn't work, and we finally give up and wait for the next aircraft.

Throughout this chapter I have tried to give you examples of how important it is to be knowledgeable about regulations, procedures, and the impact that different situations might have on these legal requirements. Controllers and pilots frequently have to make decisions, often with very little time for thought, that affect these situations, and possibly determine the legality of an operation. There are literally hundreds of words, actions, or implications that we all use on a daily basis that could amend, limit, add, or delete requirements for a given operation. These changes could drastically alter the separation standards that are used and turn what is a good, safe, and legal operation in one set of circumstances into an unsafe operation that reeks of poor judgment in another.

We could probably take most regulations in the FARs and almost every procedure in the ATC handbooks and show you ambiguities or misunderstandings similar to the ones that we have discussed in this chapter. This does not demonstrate an inadequacy in the regulations. On the contrary, these regulations are very well written considering the dynamic nature of aviation. We must be able to adjust to changing conditions and, more importantly, consider the implications of the adjustments we make. I hope that this chapter has given you a new respect for what is usually one of the driest parts of our training. Understanding the regulatory parts of your profession or avocation may take a little more effort than you thought, but the rewards are well worth the time spent.

6

Airspace

IF YOU EVER WANT TO TWIST PILOTS INTO MENTAL KNOTS, JUST TRY THIS: ASK them to define the difference between control zones and control areas, or the difference between the continental control area and the continental positive control area (a brand new airspace term). As they answer, look them directly in the eyes and listen very intently. Wait until they finish, then raise your eyebrow ever so slightly to put a quizzical, knowing look on your face and ask them, "Are you absolutely sure of that answer?"

It's kind of a nasty thing to do, but it's fun to watch them screw up their faces and think about what they have just said. Even if they get it exactly right, it doesn't require a lot of effort or double talk on your part to convince them that they didn't. The reason that it's so easy is that virtually every pilot or, for that matter, controller can relate to a personal experience where they remember having confused some aspect of controlled airspace. I have done this during a coffee break in flight instructor refresher courses as a means of slowing down instructors who are handing out information as though you absorbed it by osmosis. I use it rhetorically in advanced ground school courses to illustrate the point and, occasionally, in a flight training environment to bring a cocky student back to reality.

Imagine these same questions being asked on an FAA exam or, worse yet, by an FAA examiner in the midst of a flight check. You study all night for that exam and prepare for weeks for that flight check and then, suddenly, your mind is a total blank. Why, you ask yourself, is this information so confusing? Why are there so many different kinds of airspace and so many different weather requirements for operating in them?

Minimum VFR Visibility and Distance From Clouds				
ALTITUDE	Uncontrolled Airspace		Controlled Airspace	
	Flight Visibility	Distance From Clouds	**Flight Visibility	**Distance From Clouds
1200' or less above the surface,regardless of MSL Altitude	*1 Statute Mile	Clear of Clouds	3 Statute Miles	500' Below 1000' Above 2000' Horizontal
More than 1200' above the surface, but less than 10,000' MSL	1 Statute Mile	500' Below 1000' Above 2000' Horizontal	3 Statute Miles	500' Below 1000' Above 2000' Horizontal
More than 1200' above the surface and at or above 10,000' MSL	5 Statute Miles	1000' Below 1000' Above 1 Statute Mile Horizontal	5 Statute Miles	1000' Below 1000' Above 1 Statute Mile Horizontal

* Helicopters may operate with less than 1 mile visibility, outside controlled airspace at 1200 feet or less above the surface, provided they are operated at a speed that allows the pilot adequate opportunity to see any air traffic or obstructions in time to avoid collisions.

** In addition, when operating within a control zone beneath a ceiling, the ceiling must not be less than 1000'. If the pilot intends to land or takeoff or enter a traffic pattern within a control zone, the ground visibility must be at least 3 miles at that airport, if ground visibility is not reported at the airport, 3 miles flight visibility is required. (FAR 91.105)

Let's start by dealing with the weather requirements questions first. We will move into the airspace segments as a logical progression of how the weather requirements change in different parcels of airspace.

VISIBILITY AND CLOUD CLEARANCE

A lot of pilots think that they have to know a completely different set of visibility and cloud clearance rules for every segment of airspace in which they operate. Actually, there are only three different sets of rules that you need to know in uncontrolled airspace, and two in controlled airspace. The applicable rules change according to the altitude of the aircraft, and the points where they change are the same for both categories of airspace. Also, the visibility and cloud clearance rules for the highest altitude segment are the same for both kinds of airspace. Let's look at them a little more closely.

Note that the baseline altitude on all three rows of this chart is 1200 feet AGL and that the weather requirements change depending on whether you are at or below, or at some point above, that line. You might ask why such an odd figure as 1200 feet AGL figure was chosen as the break point. I have asked this question of several people whom I would consider to be aviation experts and have received a number of different answers.

Some say that it was chosen because of convenience and continuity. It is the altitude above which most controlled airspace begins. Others point out that this is also an altitude which is approximately 500 feet above what, years ago, was the typical airport traffic pattern environment. (Controllers normally separate VFR traffic, vertically, by 500 foot-increments from IFR traffic/airspace, so this figure makes sense when used in that context.) This allowed the traffic pattern at many airports without a control tower to be in uncontrolled airspace. Pilots could remain below and clear of clouds while in the traffic pattern without the 500-foot below-the-clouds clearance requirement. The one-mile visibility requirement permitted touch-and-go activity to continue in marginal visibility.

(Even today, at airports without instrument approaches, the 1200-foot floor of controlled airspace permits a standard 1000-foot pattern to be flown in uncontrolled airspace. But at any airport with a published instrument approach procedure (i.e., any significant airport) today's 1000-foot pattern is well above the 700-foot floor of the associated transition area, and therefore subject to controlled airspace weather minimums.)

In uncontrolled airspace *above* 1200 feet AGL, the rules change to include a requirement to remain a certain distance from clouds.

I emphasize the word *above* in this instance because of the importance of the definition of words used to describe altitude segments in the National Airspace System. When a segment of airspace is described as being *from* one altitude, let's say 1200 feet AGL, *to* another altitude, say 10,000 MSL, it means that this altitude segment begins at 1201 feet AGL and goes up to and includes the altitude of 10,000 MSL. When this top altitude is not included in an airspace segment, that segment is defined in the regulations as being *to but not including*. Another way of stating that an airspace segment does not include the exact upper altitude is, *more than 1200 feet above the surface but less than 10,000 feet MSL.*

Similarly, when defining a segment as being *from the surface to* 1200 feet AGL, that segment does not begin until your wheels leave the ground. It continues while you are operating at exactly 1200 feet AGL unless stated that it does not include that altitude. As a result, a pilot operating in VFR flight conditions at the bottom altitude of an airspace segment is considered to be below that segment.

These are rather important technical distinctions, because the difference in one foot of altitude can completely change the requirements under which a pilot has to operate. (I will elaborate on this point shortly, but for now let's stick with this topic.) Our hypothetical pilot can legally operate at 1200 feet AGL under a solid overcast which is just above the tail section of his aircraft. However, if he climbs ever so slightly, or flies over a valley which puts his aircraft at 1250 feet AGL, he would, by definition, be required to have a 500-foot cloud clearance above him. Of course he could descend 50 feet, which would place him back at the correct altitude, but he would need a radar altimeter to conduct this roller coaster flight.

This is a somewhat exaggerated example of the circumstances that most pilots face, and it is unlikely that someone will be out there with a tape measure checking on you. I use this illustration only to make the point that, if you want to operate on the ragged edge, you can expect some difficulty in maintaining the exact altitude which will keep you in the various segments of uncontrolled airspace.

There is one other thing that you should remember when you do this. If something goes wrong with the flight that calls FAA attention to your operation, there is always the possibility that you could be charged with operating your aircraft in a *careless and reckless manner*.

This particular regulation is considered so important that it is actually a two-part rule. The first part, FAR 91.9, refers to aircraft operations in general, including any airborne operations. The second regulation, FAR 91.10, is an even more specific requirement for safe operations on the surface of an airport. Remember our old friend Murphy? He loves people who cut corners.

Now, let's go back to the statement I made that there are rather technical distinctions between word usages in airspace designations and expand on this concept a little. This will open up a can of worms that should really confuse you. (I certainly don't want you to be bored and not paying attention.)

When we used the example of a pilot cruising along at 1200 feet AGL, we were discussing a situation where the pilot would have been in uncontrolled airspace on either side of that altitude. This distinction is actually important for two reasons. First, if the airspace above 1200 feet AGL had been part of a controlled airspace segment, the visibility requirements would have changed along with the cloud clearance requirements. Second, and perhaps more importantly, the exact altitude of the base of the upper layer may or may not have been included in that segment, depending on the wording of the airspace designation.

Our example was built around a pilot operating in accordance with FAR 91.105, *Basic VFR weather minimums*. This FAR defines the distinct altitude segments for basic VFR weather minimums as:

> 1200 feet or less above the surface.
>
> More than 1200 feet above the surface but less than 10,000 feet MSL.
>
> More than 1200 feet above the surface and at or above 10,000 feet MSL.

Paragraph (e) of this section states, ''For the purposes of this section [basic VFR weather minimums] an aircraft operating at the base altitude of a transition area or control area is considered to be within the airspace directly below that area.'' This means that, for example, if an aircraft is flying exactly at a 1200-foot AGL base of controlled airspace, the aircraft is considered to be below that controlled airspace, i.e., in uncontrolled airspace. It's important to note the qualifier, *''For the purposes of this section''*. This means that this concept applies *only* for the purposes of determining VFR weather minimums. It does *not* mean, for example, that when you fly at the base altitude (floor) of a TCA or ARSA you are below (outside of) that TCA or ARSA. Indeed, in such a case, you would be inside of that regulated terminal airspace, as we will discuss momentarily.

READ THE REGS CAREFULLY

It is extremely important to know exactly what defines an airspace segment, because of the existence of different requirements to operate in those segments. The FAA uses several different terms to define airspace segments. Most of these, such as *to but not including, at or above, at, at or below,* and *to and including*, are self-explanatory.

Unfortunately, there is one word which we use in airspace definitions that, when used alone, means one thing, but when used with one other word, means something entirely different. The difference between them is very subtle and the word combination could easily be read into the context of a sentence. As innocent looking as it may be, this subtle difference could mean that you would be operating legally in one case and could be in violation of FARs and have your license suspended in the other.

Read the following statement and ask yourself exactly where this airspace segment begins:

> Transition Area—Controlled airspace extending upward from 700 feet or more above the surface of the earth when designated in conjunction with an airport for which an approved instrument approach procedure has been prescribed . . .

Now let's look at more of this definition of *transition area* from the *AIM* and FAR 71.13:

> . . . or from 1200 feet or more above the surface of the earth when designated in conjunction with airway route structures or segments . . .

Compare the wording of the two statements with particular attention to the two words just prior to the altitude designation, and you will see the critical difference between the two.

When the words *upward from* are used in an airspace description, the altitude mentioned is included in the airspace segment. In this case, the altitude described means "at or above." When just the word *from* is used, that altitude is included in the next lower airspace segment. You might think that the difference in only one foot should not be a big deal, but read on and you will see where that one foot really counts.

There is one other word that keeps popping up in a discussion of airspace segments that really causes a problem for pilots and controllers. This word also has to do with the bottom of an airspace segment and is most frequently found in the regulations when we are talking about TCAs. The *base* or *designated floor* of a TCA refers to the bottom altitude in a particular airspace segment of a TCA. In the minds of a lot of pilots, there is some question as to whether this exact altitude is actually a part of the TCA. (This same wording and question also applies to the description of ARSAs.)

Not all TCAs are exactly the same size and shape, and consequently there is not a general description of the TCA altitudes in FAR 91.90, *Terminal control areas*. We do get a clue to the fact that TCA airspace segments include the base altitude by reading FAR 71.12, which is also titled *Terminal control areas*. This FAR directs us to Subpart K for a detailed description of specific TCAs, but it also states that TCAs are "controlled airspace extending *upward from* the surface or higher . . ." (The words *upward from* are not italicized in the actual regulation although they should be.) This wording places the base altitude or floor of the TCA *in* the TCA. Considering the requirements for operating in the TCA and the possible penalties for being there without a clearance, this is an important distinction.

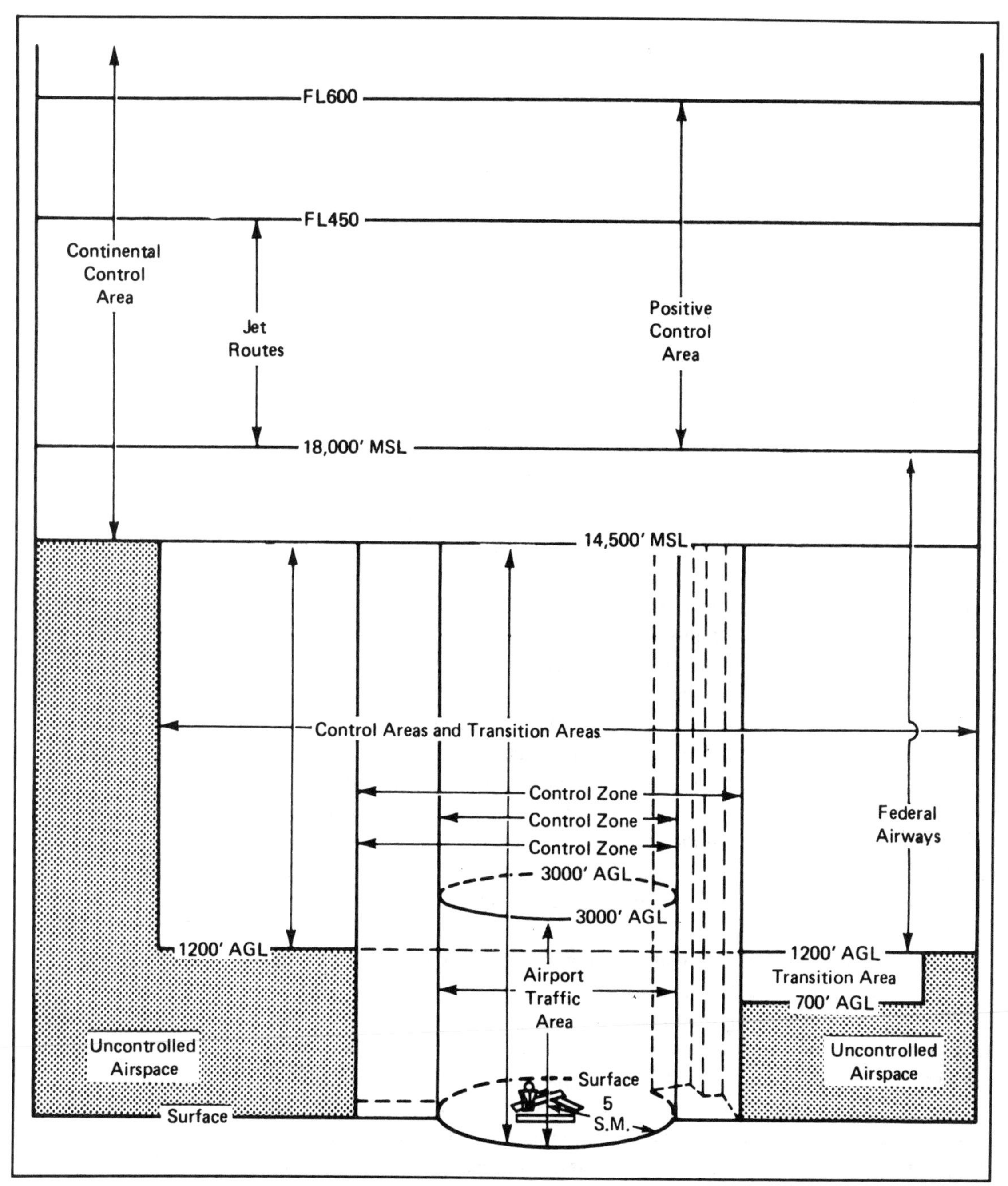

Elements of the National Airspace System. At this writing, proposals are in the works that would drastically change the names and some limits of these airspace areas.

Now that we have dabbled with airspace designations, let's elaborate on them. We will also talk about their requirements and why there is so much confusion about the different kinds of airspace.

AIRSPACE ANATOMY

If you refer to the accompanying airspace diagram, you should note that the average general aviation pilot does not often use large portions of the airspace because they rarely operate at the higher altitudes. These pilots learned the names and locations of the continental control area and positive control area years ago so that they could pass an FAA test, and much of that information was forgotten two seconds after the test monitor said, "Time is up." This lack of use eliminates the potential for reinforcing the learning and, when combined with a poor initial learning technique, leads to a very rapid loss of that information.

If forgetting can be attributed to lack of use, you would think that those pilots who operate in this environment on a daily basis would be experts on the subject. Unfortunately, this is not always the case. These commercial and military pilots are rarely required to think about the areas in which they are operating because they are controlled virtually from the moment they taxi out to the moment they stop taxiing in. In most cases these flights are conducted on an IFR flight plan, which effectively cancels the need to know the visibility and cloud clearance rules, and the average pilot is not often tested on this knowledge.

Consequently, unless an individual pilot takes the trouble to remain current on this information through his/her own studies, or is involved in a training situation where he or she is teaching the subject, the knowledge quickly becomes muddled and tends to seem much more complex than it really is.

Another reason why pilots and controllers find the National Airspace System so confusing is that they probably tried to learn the airspace as a complete package. They simply tried to memorize the entire airspace chart. While many of them can still recite the data like a tape recorder, they have trouble applying this knowledge in the real flying environment where conditions can change in a very subtle manner. Later on in this chapter, I will give you an example of how this can come about by relating a story about a temporary air traffic control tower and the airspace problems that this brought about.

This approach to learning would be like trying to understand how the human body works by memorizing the names and locations of body parts. Many pilots don't realize that each element in the airspace system is part of the building-block concept on which the National Airspace System is constructed and that they are all interdependent. Particular segments exist because the amount and/or complexity of the traffic at or between certain locations has dictated that some type of control is needed. These airspace elements exist for a reason, and some of them only exist when certain facilities are present (such as a control zone, which can only exist if a control tower, FSS, or weather observer is present). The level of control needed at a given location and the reasons for that need are what make these airspace elements different and are the fundamental reasons why so many different types of controlled airspace are needed. Let's talk a little bit about that subject.

An instructor associate of mine, who is the manager and chief flight instructor of Epps Air Service in Atlanta, Georgia, conducts the flight instructor refresher courses that I attend biennially. During the sections of these courses that deal with airspace requirements, he uses an amusing story about a ''millionaire flight instructor'' (a contradiction in terms?) as a teaching technique to illustrate how controlled airspace is constructed and why that airspace is needed.

He has graciously permitted me to use his story, so, with the exception of changing the names to protect the guilty, expanding it, and adding a few embellishments, allow me to relate that story. When we are finished, we will talk about a few specifics that most pilots and controllers always seem to get confused. (Please understand that, in this story, great liberties have been taken, especially concerning airport ownership changes and the ease and speed with which facility changes occur.)

THE SAGA OF PEGGY PIPER

Our main character is Peggy Piper, the millionaire flight instructor who probably made all of her money because her students take at least 400 hours to earn their private pilot certificate. Peggy is really tired of having to put up with all of the regulations, airspace and communication requirements, and air traffic delays associated with the complex area where she flies. She also realizes that she is burned out and her students are starting to suffer as a result (the last one took 500 hours). She yearns for the simpler life of no pressure or deadlines, no students who want to fly at the crack of dawn, and of flying in unrestricted territory where she can do what she wants, when she wants.

Peggy packs up her bags, pulls all of her money out of the local banks, which of course starts a depression in the local economy, and moves out to a little town called Nowhere, in the state of Rocky Mountain, USA. Peggy buys some property and prepares to settle down in a life of relaxed luxury. It seems that she already knew that there was a large ranch for sale in the foothills close to town, because she and one of her students got lost on a cross country trip in this area.

After a few months of living the quiet life of relaxation, Peggy has gotten back most of the drive that made her such a success, and she is looking for a new challenge. In addition to her flying skills, Peggy has one other talent that sets her apart from the run-of-the-mill flight instructor. She is a gourmet cook who can make some of the best barbecue that ever came out of a four-legged critter. So, after thinking it over for a few days, she decides to open a restaurant which she calls ''Peggy's Barbecue Kitchen.'' The new venture is a huge success and within months people are flocking to the restaurant in droves.

Some of the people who frequent this establishment are also pilots who keep telling Peggy that she should clear off some of the land behind the restaurant and put in a grass strip so that they could fly in for lunch. Since Peggy is also beginning to feel the urge to do a little flying of her own, this sounds like a very good idea. She hires a local contractor to put in a 1500-foot grass strip with a taxiway that leads right up to the restaurant. Within a very short period of time, pilots from all over the area are flying in to the restaurant for lunch and dinner. Peggy's place is doing even more business than before and she has to expand the restaurant and the airport parking areas to meet the demand.

On the Chart

Shortly after Peggy completed the airstrip, an official of the National Oceanic and Atmospheric Administration (NOAA) was studying recent satellite photos of the area and noticed that there was a small airfield on the photos that was not depicted on the sectional charts. He depicts the location of this airstrip on the next sectional chart as a magenta circle with the letter U in the center to indicate an unconfirmed airport. There are a series of dotted lines placed next to the symbol to indicate that information about the field, such as elevation, was not available. He also contacts their nearest Department Of Commerce office to have someone sent out to talk to the individual at the site of the airfield.

Several days later, Peggy notices a government car pull up to the restaurant. An official-looking government type introduces himself to her and says, "I'm from the government and I'm here to help." After a long discussion during which Peggy learns that the airport is already depicted on a sectional chart, she agrees to provide them with the information concerning the airport as long as they understand that this is a private-restricted field. She also agrees to keep the agency advised of any changes that occur on the airport by advising the FAA of the changes. Several months later a new sectional chart is published showing a magenta circle with an R in the center indicating a private-restricted airport at the location of Peggy's place and an information tag showing the field elevation and runway length.

After a while, other pilots who frequent the restaurant convince Peggy that she needs to lengthen the runway so that they can fly their twin-engine aircraft in for lunch. She expands the runway to 3000 feet and adds a UNICOM radio when two pilots have a close encounter of the dangerous kind while trying to land from different directions. Peggy dutifully notifies the FAA of the change. A short while later, the new sectional chart shows Peggy's place as a private airport with a 3000-foot strip and a UNICOM frequency. Even though the runway has been lengthened, it has not been paved, so the chart still shows a magenta circle with an R in the center.

By now the restaurant is beginning to cater to local business meetings and small conventions which run late into the evening. The pilots begin to complain to Peggy that the airport needs to be equipped with lights so that they can operate safely.

This sounds like a good idea to her since everything that she has done so far has always increased business at the restaurant. Peggy adds the lights and notifies the FAA. On the next chart the information tag associated with Peggy's place has an L indicating runway lights. Since she also added a rotating beacon on the airport, the airport symbol also has a star on top.

Open to the Public

A short time passes during which Peggy continues to prosper and make changes to the business she runs. Her customers convince her to add some fuel tanks so that they can refuel and a small maintenance facility so that they wouldn't have to bring in a mechanic to fix their broken airplanes. Then a couple of local entrepreneurs, who are anxious to take advantage of Peggy's reputation as a pilot and the volume of business at the airport, suggest that she allow them to make further improvements at the airport. They want her to go into business with them by opening a flight training facility and charter service under her name and in conjunction with the existing maintenance and fuel facilities. They also ask her to allow the facility to become a public access airport.

Peggy agrees, and Peggy Aviation is born on an airport that has been expanded to a 5000-foot paved runway, hangar facilities, full maintenance and fuel services, and a limousine company to carry people back and forth to the restaurant which now has a large motel situated nearby. The motel came about because the pilots needed a place to stay while their airplanes were being fixed. The state of Rocky Mountain has also given the airport a name, Peggy County Airport, and an identifier RM31. It seems that a state official, grateful for Peggy's support in the last election, has pulled a few strings during the state's most recent political reorganization. The name, Peggy County, was given to the million square acres that our flight instructor owns.

Peggy notifies the FAA of the improvements and of the fact that the airport is now open to the general public. The next chart shows the airport location using a solid magenta circle which contains a symbol for a paved runway over 1500 feet laid out in the direction that the runway is aligned. The magenta circle also has four little "ticks" attached to the top, bottom, and sides to indicate fuel and maintenance facilities at the location. It still has an information tag that lists runway length, lights, and UNICOM frequency. The airport is now also listed in the *Airport/Facility Directory (A/FD)* with a complete list of facilities and services.

The pilot requirements for operating into this airport up to this point have not changed. When flying up to 1200 feet AGL, pilots must remain clear of clouds and have one mile flight visibility. During that portion of their flight to Peggy's place that occurs above 1200 feet AGL but less than 10,000 feet MSL, they still must have one mile flight visibility but are required to remain at least 500 feet below, 1000 feet above, and 2000 feet horizontally from clouds. Even though there is a UNICOM on the field, there is still no requirement, other than the rule about careless and reckless operation of an aircraft, to use communication equipment.

Adding Approaches

Aircraft traffic in and out of the airport, restaurant, flight school, and motel have by now increased to the point that Peggy's old political friend, who is now a U.S. Congressman, suggests that the field should have an instrument approach so that people could get into the airport during bad weather. He points out that the flight school, which is now the largest in the state of Rocky Mountain, could use the equipment for practice approaches. It seems that the students were having to fly across the state line to do this and were spending valuable tax money in the enemy camp.

So, Peggy purchases a terminal VOR (TVOR) to be installed on the airport and uses her contacts with the government to have VOR approaches designed and flight checked for both runways. She knows that later on she will want to add an ILS system to the equipment at the airport and, because she can get a good package price, she also buys a nondirectional beacon (NDB) to be installed 5.5 miles from the airport along the extended centerline of the runway.

When the FAA arrives in the area to check out the two VOR approaches, they also flight check the NDB for an NDB approach.

Meanwhile, FAA airspace specialists establish a bypass Victor airway connecting the Peggy County VOR (which is assigned the new airport identifier, PCA) and an existing high-altitude VOR less than 25 miles away. (TVORs like PCA are considered to be reliable for use as navigation aids up to 25 miles and 12,000 feet. These altitude and range boundaries are called the *service volume*.) The Victor airway runs directly over the Peggy County Airport.

While these are small, subtle changes in the NAS, they bring about some significant changes in the airspace system and the requirements to operate at Peggy County Airport. A *transition area* is established in conjunction with instrument approaches. On the new sectional, this is shown as a magenta-colored, keyhole-shaped area surrounding the airport. In this area, controlled airspace now begins at only 700 feet AGL. Elsewhere near the airport, controlled airspace starts at 1200 feet AGL, indicated by the blue tint. Below 10,000 feet MSL, the minimum VFR visibility and cloud clearance requirements in these airspace areas are now 3 statute miles flight visibility, and cloud clearance of 500 feet below, 1000 feet above, and 2000 feet horizontally from clouds. Since there is still no certificated weather reporting service available at the airport, these determinations must continue to be made by the pilots operating in the system.

The new chart also indicates the location and frequency of the NDB, and a compass rose symbol and information tag for the new VOR. These changes are also indicated in the *A/FD*. Concurrently, the three new instrument approach procedures are published by National Ocean Service and Jeppesen.

Time for a Tower

Already there have been several near-misses, including one involving the Congressman, who convinces Peggy that the level of traffic justifies a control tower. During their discussions, he also informs her that the FAA is going to locate one of its new automated FSS facilities somewhere in the region and that he would like to see it built at Peggy County Airport. (Peggy deeded the runways and taxiway areas to the county when she put in the VOR and now serves as airport manager only.) He says that this is a particularly difficult time for getting all of this done as this is an election year and he is running for a seat in the U.S. Senate. Since he supported spending reductions and is currently campaigning on fiscal responsibility, it would be unwise to work toward the establishment of a big federal facility at this time. He suggests that the county install a nonfederal control tower at the field now and he will work to get it converted after the election.

Peggy is suitably impressed with his arguments and within a few months the new nonfederal (county-operated) control tower is operating during daylight hours. This brings about a change to the color of the airport symbol on the sectional chart. The new charts now show a blue circle replacing the old magenta circle which indicated an uncontrolled airport, The blue symbol indicates an airport traffic area exists when the tower is open.

This latest change seems like it would bring about a major change in the requirements to operate in and out of the field, but it does not. The weather requirements remain the same as they were before and the only change to the communication requirements is that, when you are operating at an airport with a nonfederal control tower, you must maintain two-way radio communication with the tower if your equipment allows. If your equipment only allows reception capability, you must monitor the frequency and you must obtain and comply with ATC instructions on airport movement areas. By implication, if your aircraft is not equipped with a radio, you could enter the airport traffic area and look for light signals when you are inbound to land. On an outbound operation, you could obtain telephone approval and use light signals for operating on the movement areas on the ground.

The establishment of a tower and its associated frequency also brings about one subtle change to the communication procedures on the airport when the tower is closed. Prior to the opening of the control tower, the *common traffic advisory frequency* (CTAF), which is the frequency designed for traffic advisories, had been the UNICOM. Now that there's a tower frequency, that frequency continues to be the CTAF even after the control tower closes for the night. When the tower is closed, pilots should use this frequency for traffic information the same way that they would use the UNICOM at a non-tower airport.

The approach control, FSS, or ARTCC facility that serves this airport when the tower is open frequently has the ability to transmit and receive on the tower frequency after the tower closes. This allows the pilots to obtain or cancel an IFR flight plan on the ground, obtain nearby weather information and altimeter settings, and learn about known or observed traffic.

While all of these changes have been going on, Peggy has decided that the profit potential of adding fresh seafood to her restaurant menu would be enormous, so she sets out to locate an adequate supply. Her suppliers tell her that the fish would rot if shipped from the coast by truck. They tell her that they could fly the products into the local airport, but the runway is not long enough to handle all of the 747 cargo jets needed to haul the amounts that she will buy. Reluctantly, the county agrees to extend the runway to 8000 feet and install the equipment necessary to service this type of aircraft. These changes are noted on the information tag on the sectional chart and in the *A/FD*, but they do not affect the airport symbol on the charts. The existing blue circle with the runway alignment depicted in the center is used for airports with runways 8000 feet or less.

Enter the Feds

This latest series of changes is all that Peggy's political friend, now a U.S. Senator, needs to convince the FAA that Peggy County is an up-and-coming airport and an ideal location for their newest FSS facility. He has decided that it will be easier to convert a nonfederal tower facility over to a federal facility once there is another FAA organization already on the field. The fact that Peggy is a friend of aviation and has donated the land required for the facility also helps keep the cost of the operation within limits.

Construction of the new facility is typically swift and the next series of charts shows the airport as having an automated FSS facility with an information tag showing the appropriate frequencies. Also, the frequency of the ATIS, which the tower now has, is added to the airport information tag. The appropriate changes are, of course, listed in the *A/FD* and on instrument approach plates. Peggy is proud of what has been accomplished but she is beginning to feel that the walls are closing in on her again. What she doesn't know is that a series of unrelated events are shortly going to change the picture even more.

First, officials at the FAA are faced with a situation where they are losing the lease on some land where a long-range radar site, a microwave repeater, and a VOR are located. It seems that the owner wants to sell the land to a developer for a vacation resort and a golf course. Because of the overwhelming logic of the idea, and the fact that Peggy's friend, now the chairman of the aviation subcommittee in the U.S. Senate, has convinced them that his district would enthusiastically support the idea, the FAA has decided to relocate the VOR facility to PCA where they already own all of the land that they will ever need. They move the equipment, upgrade it to a newer, solid-state, second-generation VOR with remote maintenance capability, and redesign the airspace configurations associated with the old VOR.

Since they are also faced with having to move their radar site, they will lose most of the low-altitude radar coverage in the area. This loss, combined with the significant increase in instrument approach activity at the airport and the increased requests for radar service in the vicinity, make obvious the need for radar coverage of the Peggy County area. The FAA planners begin to consider the relocation of the radar equipment to the Peggy County site and the opening of a terminal radar approach control (TRACON) facility at the tower location.

Second, the air cargo organization that has been flying into Peggy County has been purchased by a major airline which is contemplating establishing a shuttle service into the airport to test the market. Their primary concerns about the airport are the lack of a control zone at the airport, the traffic that could be legally operating in the area under the transition area with one-mile flight visibility, and the fact that the control tower will not be open at night. They believe that the establishment of a control zone would provide more positive control of traffic in the airport vicinity and reduce the possibility of unknown aircraft operating with limited flight visibility near the approach courses and traffic patterns. They feel that the control tower should be a full-time federal facility to provide an extra margin of safety to their nighttime cargo flights. They know that the current part-time tower might not be open if their late-evening passenger flight encounters air traffic delays enroute to Peggy County.

Third, in conjunction with the opening of Peggy's expanded, full-service restaurant, motel, and dude ranch (she's still coming up with new ideas), she has won the right to host the next convention of the International Association of Newspaper and Magazine Publishers.

Meetings are held, opinions are heard, decisions are made, and when the new radar approach control facility opens, the charts depict a new control zone around the blue airport symbol. This control zone also contains extensions to protect the revised instrument approaches. The new charts contain expanded tables and information tags with data concerning the new approach control frequencies and services available at Peggy County.

Time for a TRSA

Even though few terminal radar service areas (TRSAs) are being established at new facilities, the FAA decides to implement one at Peggy County because of the level of traffic anticipated in the short term.

Finally, on the charts, the new NDBs associated with the ILS systems are depicted, and new airways served by the relocated VOR are shown going in virtually every direction. These airways have expanded the control areas which are depicted by blue shading and begin at 1200 feet AGL outside the control zone. There are several new transition area extensions associated with the instrument approach courses on the airport. Like the preexisting transition area, these areas begin at 700 feet AGL outside the control zone. Overlying all of this controlled airspace is the positive control area, which begins at 18,000 feet MSL, and the continental control area, which begins at 14,500 feet MSL except in the mountains near Nowhere (elevation 14,100 feet) where it starts at 15,600 feet MSL (1500 feet AGL).

Add an ARSA

As luck would have it, the publishers convention coincides with the opening of the new FAA facility, and a few days later, almost every newspaper and magazine in the world starts heralding this previously undiscovered paradise with its magnificent restaurant, fabulous golf course (Peggy bought out the developer), four-star motel, and low-cost vacation resort. Within weeks the new ATC facility is swamped with traffic, more controllers have to be rushed in, and the FAA has to institute an airport radar service area (ARSA). A hearing is held concerning the implementation of this new controlled airspace. The local pilots protest, but the ARSA appears on the next sectional chart.

Transition to a TCA

The four major airlines (the other three couldn't allow Tandem Airlines to establish a foothold without some competition) that have been scheduling once-a-week shuttle flights into PCA, note a significant increase in the passenger loading on their flights into PCA. All of them find that they will have to add several new flights a day just to keep up with the local traffic. The airline officials realize that the location of this airport is ideal for a major hub as long as this traffic level can be sustained. They all quietly conduct demographic and economic development studies which indicate that this is going to be the case.

The presidents of the airlines all meet with the members of the Peggy County Airport Authority (Peggy turned management of the airport over to civic leaders when it became too much work for her and now she's just a member of the board of directors) and together they decide to extend the existing runway to 12,000 feet, build a parallel 12,000-foot runway, and add an airline terminal with 200 gates. They also add a small commuter runway and terminal. Additionally, one of the airlines recently merged with a west coast regional air carrier and they decide to build a regional office complex and major maintenance facility on the airport. These changes mean that the airport symbol on the sectional charts goes from the blue circle to a full outline of the airport layout without the blue circle.

As this is being completed, airline representatives, members of the airport authority, and FAA officials meet frequently to study the air traffic situation. The existing traffic has increased so drastically that the airport already qualifies for a TCA. Another hearing is held, the local pilots protest, and the TCA is implemented. Tandem Airlines has started direct service from Peggy County to the Orient, requiring the airport identifier to be prefixed by the letter *K*, which reflects an international airport and meets the requirements of the International Civil Aviation Organization (ICAO). Our old friend, now the Senate majority leader, recognizes the problems that this will cause and sees to it that a U.S. Customs facility is built at the airport.

The sectional chart and *A/FD* update information concerning the facilities at the airport. (By now several pages in the *A/FD* are dedicated to KPCA), and a TCA chart is issued. A NOTAM is also issued about a temporary flight restriction around the convention center at Peggy's resort complex. It seems that Peggy has won the right to hold this year's presidential convention as a favor to her old political ally.

Military Invasion

Ever since Peggy turned over the day-to-day management of the airport to others, she has had some free time available. As is typical with free time, where there is a void, something comes along to fill it.

An Air Force base located near Peggy County has been taking advantage of the increased air traffic control facilities in the area to conduct pilot and instrument approach procedures training at the field. The base commander has also been giving Peggy frequent training and familiarization flights in the new advanced fighters while they try to convince her that the old hangar area on the airport would be an ideal rapid deployment facility where they could locate a newly created Air National Guard wing. He suggests that this new organization would need an adjutant who would be a Lieutenant Colonel, and he reminds her that she still holds a reserve commission in the Air Force.

As a loyal American and an old fighter jock who never got beyond her silver bars, Peggy realizes the need for scattering forces over a large area, and she agrees to donate the old facilities to the Air National Guard. The lure of the wild blue yonder and the fact that this is a huge tax writeoff helps her decision along just a little.

The next series of charts now depict several recently established visual (VR) and instrument (IR) military training routes, military operations areas (MOAs), and a restricted area associated with the military practice activities.

Politics and Mother Nature

While Peggy is working with the military on their newest project, other events are occurring that will add still more pieces to the complex airspace puzzle around Peggy County. The Senate majority leader and his family have returned home to the mountains above the airport to rest and await the election results from their recent national campaign. The senator's wife, who is also the honorary chairperson of the National Audubon Society, is taking a walk in the hills with some of the local society members when they discover a nest containing three eggs and a mating pair of three-winged eaglets. This species was considered to be extinct in the wild so the discovery is particularly exciting.

While she is observing the nest she notices several small aircraft conducting low-altitude practice maneuvers over the area, and she becomes very concerned about the safety of the birds. The senator's wife rushes home and insists that her husband immediately contact the Department of the Interior and ask them to establish a wildlife refuge area around the nesting place. The senator's reputation for being fast on his feet is well deserved, and he quickly has the right people working on the project. The next sectional chart is published depicting a wildlife refuge area. It was defined by a blue outline combined with a series of blue dots around the refuge area. A note requesting that pilots maintain a 2000-foot AGL minimum altitude over the area is contained in the margin of the chart. At about the same time, officials of the Secret Service work on establishing a prohibited area in this same area, which is now the home of the new Vice President-elect.

Ban the Students

All of these changes, combined with projected traffic increases due to airport expansion (including additional airline flights when the remainder of the terminal spaces open in a few weeks), make it apparent that traffic levels will be pushed up to the point that PCA will be one of the busiest airports in the country.

Meetings are held by all the concerned parties and together they study the facts before them. The existing and projected traffic situation, the safety requirements, the expansion of other small airports in the vicinity, and the military's plans to expand the Air National Guard base all indicate that student pilots should be prohibited from landings and takeoffs at Peggy County. This puts PCA in the same category as O'Hare, Logan, LAX, and nine other terminals.

Seems Like Only Yesterday

Several months later, the new Vice President of the United States, fresh from several days of inaugural balls, flies home in Air Force 2 to the airport that he had helped name during his days as a state official. He finds his loyal constituent, Peggy Piper, sitting in her office at the restaurant, staring out the window over the fifth busiest airport in the world. She holds the Peggy County TCA chart in her hand and has a faraway look in her eyes. The chart looks like a brown-magenta-and-blue nightmare—covered with complex control zones, prohibited, restricted, and reserved airspace of all kinds, and brimming with information tags of every description.

Peggy looks at the chart and remembers the days when she could hop into her little TriPacer and go zipping around the scrub brush and foothills totally alone. Now, the equipment she needs in the airplane just to fly in the controlled airspace around her ranch costs almost as much as what she paid for the aircraft. Behind her, scattered over the top of her desk, are travel brochures of Alaska.

The Moral to the Story

Our saga of Peggy Piper serves to illustrate how controlled airspace develops. Without airports, control towers, instrument approach procedures, VOR airways, and various levels of traffic, controlled airspace would not be needed. Install an airport with an instrument approach and you have a transition area for the instrument approach. Build a tower and you have an airport traffic area—and almost certainly a control zone will follow. Add a navigation facility nearby and you suddenly have control areas in conjunction with VOR routes (not to mention the continental control area and continental positive control area which overlie almost all of the U.S.). Add some traffic volume to the picture and you suddenly have ARSAs and TCAs to contend with.

But this story is fiction, isn't it? Real airports don't just spring up out of nowhere (no pun intended), do they? Well, while this exact scenario is an exaggerated one, just consider a few recent cases. In the early 1970s, the Dallas-Fort Worth airport was flat Texas land. The only thing flying was tumbleweed. The airport, as we know it today, was only a gleam in the eyes of some airport authority members. Much the same kind of expansion would have happened at the Florida Everglades airport had that facility been completed as planned. The Orlando airport and surrounding areas have never been their same quiet selves since some cartoon characters started prancing around a little park nearby. These are just a few examples of how the migration patterns of the American people can, and frequently will, drastically change the face of the land and the airports on that land. Where the people go, so goes aviation.

Now let's take a more in-depth look at exactly what is involved in various airspace segments and perhaps punch a few holes in some myths.

TALE OF THE PHANTOM CONTROL ZONE

Earlier in this chapter we talked about the sometimes subtle differences in airspace segments and the confusion that could result from getting them mixed up. Most of us tend to think of only two kinds of airspace when we think of this subject at all, and we allow this simple approach to limit our understanding of the real differences in airspace where we operate. Perhaps the best example of this, and a classic case of where airspace is not just defined as controlled or uncontrolled, was related to me a few years ago by a pilot in Raleigh, North Carolina.

Unfortunately, I cannot remember who the storyteller was, because I heard it while a bunch of us were swapping war stories after a pilot/controller forum. I assume that this story is true mostly because it's too far-fetched not to be. But, in any event, it serves to illustrate that everyone, pilots and controllers included, can have a problem understanding all of the rules and definitions regarding airspace designation.

It seems that a pilot was enroute to Kitty Hawk, North Carolina, to attend a fly-in that was being held to celebrate the anniversary of the Wright brothers' first flight. This was expected to be a popular, and consequently, very busy weekend, so the FAA established a temporary control tower at the airport to handle the traffic load and increase safety.

When the pilot arrived in the area, he called the controllers on the appropriate frequency and advised them that he was inbound to land at the airport. This flight was being conducted below 1200 feet AGL on a typical summer day along the Carolina coast, and the haze had reduced the visibility on the ground to about 2 miles. (The airfield had no instrument approaches and therefore no 700-foot transition area.) The controllers had studied the local area in preparation for their duty at the temporary tower and knew all of the runway lengths and the distances from the tower to various locations on the airport. When this pilot called for landing instructions they could only see about 2 miles in any direction. Since they performed this duty every day at their permanent location (which had a control zone), they responded that the field was IFR and requested the pilot's intentions.

The pilot was just about to ask for a Special VFR clearance when he realized that the airport had a temporary tower, no FSS, and no official weather observer. For these reasons, a control zone would not exist at the facility. So the control zone rules requiring a minimum of 1000-foot ceiling and 3-mile visibility would not apply. The pilot advised the controllers of this situation and of the fact that he was clear of clouds (there were none), had more than one mile visibility, was landing at an airport within the ATA, and was operating in accordance with the FAR requiring communication with the control tower. All he wanted from the controllers was an authorization to operate in the ATA.

I won't go into all of the unpleasant details of this misunderstanding, other than to say that the controllers were extremely embarrassed and apologetic when they eventually realized their error. I am reasonably sure that this was the first time these controllers had ever been confronted with this type of an operation.

My own situation is probably typical of the average air traffic controller's background with respect to control zones. Every airport where I have ever worked, flown into or out of, that had a control tower, also had a control zone. But airport traffic areas do not *automatically* come equipped with control zones. These are federally designated areas which require that a certificated weather observer, who is certified for the particular location at which he or she is operating, take observations during the time that the control zone is operational. Control zones must also be depicted on aviation charts with times of operation published if the control zone is less than a 24-hour operation.

ATAs enjoy a rather unique designation in that they are described as *Other Airspace Areas* (check the *AIM*). They are not considered controlled airspace in the sense that they are not necessarily supported by navigational aids, weather reporting facilities, radar, or instrument procedures. Pilots flying in ATAs are not necessarily subject to the same cloud clearance and visibility requirements found in controlled airspace (unless they are flying high enough to be in a transition area or other control area, or the airport also has a control zone). ATAs also do not fit into the category of uncontrolled airspace because they are specifically designated segments of airspace, and communication requirements exist for operations within them.

If professional pilots and air traffic controllers have to think about this type of situation, consider what the average student pilot must have to go through and the mistakes that he or she would make in this situation. If this story is really true, and I have no reason to doubt its validity, I wonder how many pilots were cleared Special VFR into a nonexistent control zone on that summer day in Kitty Hawk. Worse yet, I wonder how many turned around and went home because the airport, which had a control tower but no weather observation facilities, was declared IFR.

Let's analyze some other airspace segments individually and point out some misunderstandings about each. Let's begin with one that, in this age of full-time air traffic control towers and dwindling FSS locations, few pilots have occasion to use on a regular basis, the *airport advisory area*.

AIRPORT ADVISORY AREA

The *airport advisory area* is a segment of airspace within 10 miles (some FAA documents say statute miles, others imply nautical miles; at this writing, even FAA management in Washington isn't sure, but it's a minor discrepancy nevertheless) of an airport where a flight service station is operating and there is no control tower or the tower is closed. The FSS specialists at these locations will provide airport advisory services (AAS) to arriving and departing aircraft by giving them pertinent information about winds, weather, runways in use, known traffic, and any information that might impact the operation of your aircraft. The use of these services is not mandatory on the part of the pilot but, as with all FAA services, it is strongly recommended that all pilots use these services. It does absolutely no good to hurry into an airport without finding out about other traffic (thinking that you are saving time), and then have to go around because someone is on the runway or taking off in the other direction. (Be advised, control zones usually exist at airports where a FSS is located, even when there is no operating control tower. Special VFR operations within the control zone must be conducted in accordance with normal SVFR procedures.)

There is one other potential problem that occurs at some airports with airport advisory service. If such an airport also has a control tower that closes for some portion of the night, the frequency normally assigned to the tower is usually taken over by the FSS facility and operated as a CTAF. If you remember our earlier discussions about CTAFs, you know that a CTAF is not used to exercise air traffic *control*.

Pilots occasionally call on the tower frequency after the tower has closed and do not pay close attention to what the FSS specialist says in response. They are accustomed to hearing weather and runway information from the tower controller, and that is what they are listening for when they call on the tower frequency. They fail to note that the FSS specialist answers as "XYZ Radio" and not "XYZ Tower" when he or she answers. They also tend to miss the other key elements in what the specialist says as being different from standard air traffic control *instructions*. Having "heard" what they initially expect to hear, they continue inbound to the airport and are surprised and sometimes vocally abusive to the specialist when the remainder of the services that they expect from a tower controller are not forthcoming. This is particularly true when they encounter other aircraft and the FSS specialist has not given them a sequence or advised them to adjust their flight path to follow a particular aircraft.

Even though a control *zone* might continue to exist after the tower closes on those fields with FSS facilities, traffic is no longer *controlled*. Most pilots are taught this fact early in their training but they develop expectations when operating into airports equipped with control towers. Unfortunately, these expectations usually don't have the same hours of operation as the control tower. Pilots need to pay particular attention to tower operating hours and they must *listen* to what is said in response to their call to an ATC facility.

MILITARY OPERATIONS

Perhaps you have noticed that our discussion of pilot errors seems to center repeatedly around infrequently used procedures or situations where pilots are not as familiar as they should be with the types of operations going on around them. An example of an operation that meets both of these criteria, and one which is conducted in another series of airspace segments, is military operations activity.

Like airport traffic areas and airport advisory areas, the various military activity airspace is not specifically defined as either controlled or uncontrolled airspace. I will leave it to you to study the various airspace segments in detail, because an in-depth study of these areas alone could easily constitute an entire book. But let's review them in general and point out that they do have a place in the general scheme of aviation.

Military activity areas are types of *special use airspace* that are specifically designated and charted areas (with the exception of *controlled firing areas* which are not depicted on aeronautical charts) within which various kinds of military training and practice are conducted. These *restricted areas, warning areas, military operations areas (MOAs), alert areas* and *controlled firing areas* cover the majority of military training activities that occur on a continuing basis. I should point out that alert areas are not necessarily limited to military activity. For example, there is, or at least I seem to remember that there used to be, an alert area associated with the Daytona Beach airport area. This alert area was brought about by the high volume of student training traffic generated by the Embry Riddle Aeronautical University and other flight training facilities in the area. I'm sure there are other such areas but I am not personally familiar with them.

It is usually not difficult to navigate around these areas and many pilots find that a simple communication with the controlling agency can often result in the ability to fly through these areas when they are not in use. The controlled firing areas pose even less of a delay to traffic because activity in these areas is terminated immediately when an unknown target appears on radar or a spotter observes an aircraft approaching that is not participating in the activity. Many pilots have flown through these areas without even knowing that they existed.

One military activity that does occur outside of these designated special use airspace areas and does have the potential to impact your operation, can be found described in the *AIM* airspace section titled *Other Airspace Areas*. These *military training routes* (MTRs) are IFR or VFR high-speed, low-level training routes designed to give our airborne military forces some low-level combat and terrain-following training experience. These are done outside of military reservations to simulate how they operate against a real enemy.

The key point that I want to make in this discussion is this: I want you to ask yourself if you know (1) where these routes are in the airspace where you normally operate, and (2) how to find them on the charts if you travel to an unfamiliar area. If you have been flying for any length of time, you probably have heard a few pilot war stories about close encounters of the military kind. The military pilots in these situations probably had the civilian targets on their radarscopes, and there was probably less chance of a collision between those two than between two general aviation aircraft. But this is not always the case, and even if it is, the civilian pilot is no less shocked when this high-speed projectile goes buzzing by. Of course we all know that U.S. military pilots would never practice intercepts against innocent civilian targets.

IR routes (IFR MTRs which are normally conducted at 1500 AGL) and *VR* routes (VFR MTRs which are normally conducted below 1500 AGL) are depicted as grey lines on your sectional charts. They have an arrow, next to the numbered name of the route, that indicates the direction that aircraft will be following along these routes. There are two important facts to note in connection with these routes that may not be immediately obvious. First, just because you see an arrow indicating that traffic will be traveling in one direction on that route, don't be surprised to find traffic going the other way. A close look at one of these routes will frequently show an IR or VR route with an arrow going in one direction at one point on the grey line, and an entirely different IR or VR route number with an arrow going in the other direction somewhere else on that same line.

Second, pay particular attention when operating in the area on the chart where one of these lines comes to an end. This is usually the area where aircraft are either entering or leaving the training route and they may be descending into or climbing out of the standard operating altitudes at these points (SEE CHART EXCERPT). In any event, use particular caution when operating anywhere near these areas, because the pilots operating the aircraft on these routes are just as subject to making a mistake as any other human being.

You should remember *all* of these military areas. They have a sneaky ability to fall into the category of being out of sight, out of mind. Mr. Murphy has a district office near every one of these activity areas and I am personally familiar

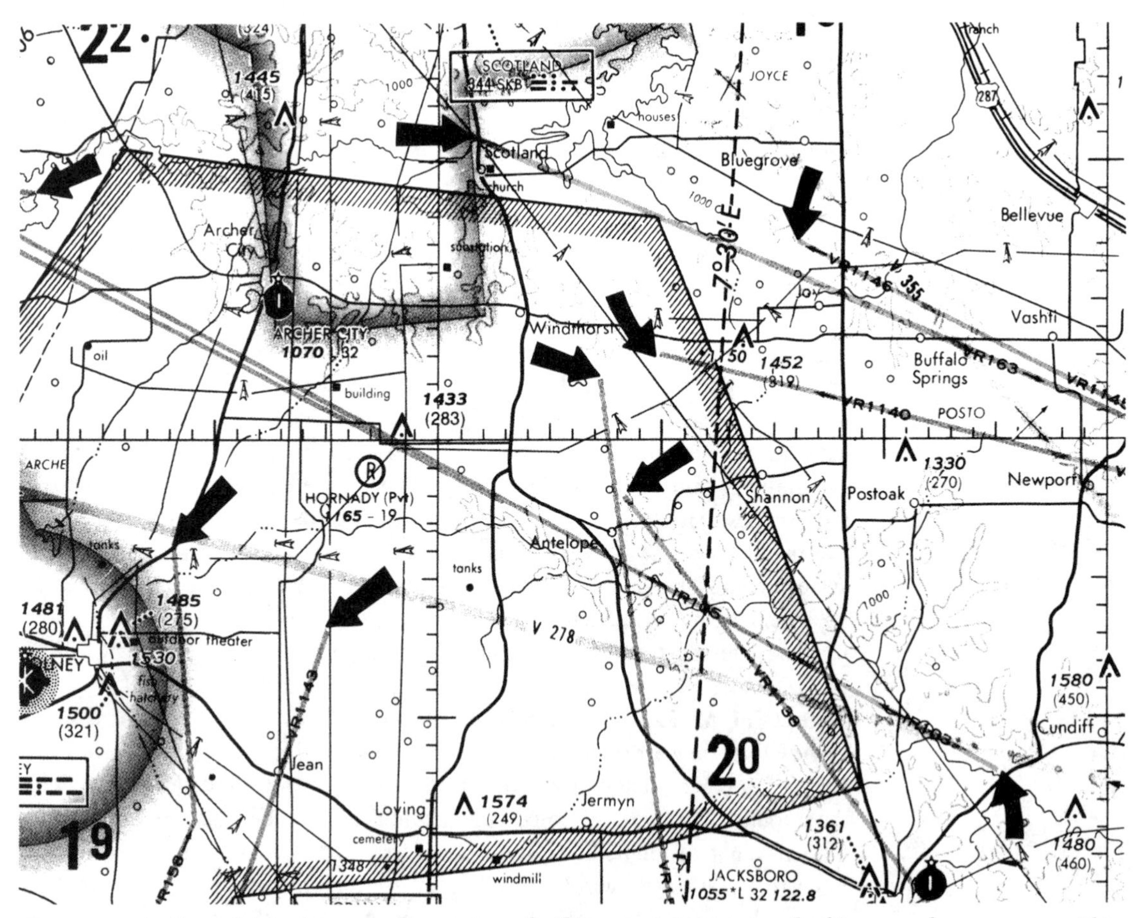

The arrows indicate beginning or ending points of military training routes. In this example, some are within a military operations area (MOA) and others are not. Use extra caution at these locations. They are not considered safe places for civilian practice areas.

with several scary episodes connected with them. One pilot I know had a 20MM hole shot in his wing. He was lucky that the cannon shell did not explode because it would have taken off his wing if it had. Another pilot, who was a flight instructor who should have known better, chose a practice area that had a VR route running right through the middle. His student quit flying after the close call with an F4. He is lucky that he doesn't have to live with the thought that someone died because of his negligence, but it's bad enough that a potentially good pilot will never fly again and will never know the pleasures of piloting an aircraft.

Now, let's move on to a discussion of an airspace segment that we have already talked about in some detail, the terminal control area.

"REMAIN CLEAR OF THE TCA"

Perhaps the most important requirement for operating in a TCA and the one most frequently violated and/or misunderstood by pilots is the rule requiring them to receive "authorization from ATC prior to operation" in a TCA. The fact that a pilot has established contact with the controlling agency does not constitute such an authorization, nor does radar identification by that agency automatically provide such an authorization. Even though it is technically not required, most controllers have developed the habit of saying "remain clear of the TCA" to almost every initial contact with an unidentified aircraft.

We have found over the years that pilots will cruise along into a TCA given the slightest indication that they have permission to do so. Frequently, that permission exists strictly in their own minds, and they have been told by the controller and by regulations that they are not allowed to do so. Let's take a look at a typical transmission between a pilot and a controller, and I will show you what I mean. In this situation we will take a VFR pilot who is operating at 6500 feet MSL, under an 8000-foot shelf of a TCA on a course that will take him to a satellite airport that is under a segment of the TCA whose base is at 3500 feet MSL:

Pilot: *Big City Approach, this is N123TA over Phantom Lake inbound to Smallville with Information Alpha, requesting vectors to the airport.*

Controller: N123TA, Big City Approach, remain clear of the Big City TCA. Smallville landing Runway Three Five, their altimeter 2996. Squawk 0336 and verify your current altitude.

Pilot: *N3TA squawking 0336, and we are level at 6500.*

Controller: N3TA, radar contact two five miles southeast of the Big City airport, suggest heading 350 to the Smallville airport.

Now, given the above exchange between the pilot and the controller, ask yourself if this pilot is authorized to enter the Big City TCA. The answer, of course, is that he has not been given such a specific authorization. In fact, in addition to the regulatory requirement to remain clear of the TCA until he has obtained a clearance, the controller has told him to remain clear of the TCA. Neither the establishment of radar contact, nor the suggested heading to his destination airport constitutes such an authorization. Yet pilots will go so far as to hire a lawyer to fight the fact that we take exception to their blasting through the TCA in such circumstances. In situations like the one I have just described, we hear statements like, "If the controller had wanted me to remain clear of the TCA, he wouldn't have given me a heading to fly," or "The controller wanted me to remain at 6500 feet because he verified my altitude and never gave me a descent."

It happens almost every day. But there is another side to this story of TCA operations. It is often not the pilot that is operating illegally that gives us the most cause for concern. The top of the Atlanta TCA is 12,500 MSL and it is not unusual for us to see an aircraft cruising along at 12,600 MSL along a busy arrival or departure route. The traffic we control into, out of, and over our satellite airports notwithstanding, we move upwards of 800,000 aircraft a year into and out of the Atlanta airport. This means about one aircraft every 40 seconds (26 seconds if we use a more realistic 16 hour day). It is quite likely that these aircraft are going to have to climb or descend through the altitude that this pilot is occupying. In most cases, this aircraft is at that altitude without talking to anyone. I often wonder what the busy executive in the back of that aircraft would say if he or she knew what the pilot was doing.

We usually stop our climb and/or descent at an altitude sufficient to protect the unknown pilot in this situation. We also take into consideration the fact that the Mode C in that aircraft may legally be off by as much as 300 feet when we assign our aircraft an altitude to miss the traffic. Sometimes though, our ability to assign an altitude to miss traffic is limited by the course that the aircraft must fly to conduct its operation.

As much as pilots like to skirt along barely above the top of a TCA, they seem to enjoy occupying the airspace 100 feet below the bottom of a TCA even more. If you were to watch that part of the Atlanta TCA along the final approach courses where the base of the TCA drops to 2500 or 3500 feet MSL, you would see a steady stream of traffic buzzing along at 2400 and 3400 feet respectively.

I have often watched the radarscope with a considerable amount of concern as I see two aircraft flying the same DME arc around the edge of the TCA in opposite directions at the same altitude. They are both very familiar with the lateral and vertical limits of the TCA, but they were playing hooky from school when the physics teacher explained that no two objects could occupy the same space at the same time. Since we are usually not talking to either aircraft, we can just watch and hope that the Big Sky Theory works again. I know that some of these pilots have gotten a big scare in these situations because, if you watch the flight paths after some of these targets cross, you will note some rather erratic tracks for a while before they reestablish on course. But the old adage that a ''miss is as good as a mile'' is not necessarily true in this situation.

I mentioned that we sometimes cannot assign an altitude that will ensure that we will provide enough room to protect these unknown aircraft. This is not because we don't try, it's simply that some conditions exist that preclude that possibility. Many of the aircraft that we see skirting along below the TCA are doing so directly beneath the final approach courses into Atlanta. We issue this traffic to the aircraft on final and if necessary we will move one of our aircraft if the situation is critical. But the real danger to these zoomers comes from the wake turbulence generated by these large, heavy aircraft.

We are required to provide at least 5 miles or 1000 feet vertical separation to an aircraft that is operating behind and below a heavy jet aircraft. Pilots are supposed to know this and yet we routinely see targets, whose speeds indicate that they are small lightweight aircraft, cross directly behind and under one of these big monsters with less than a couple hundred feet of vertical separation. Certainly the pilot sees this aircraft and thinks that he or she is maintaining visual separation when they do this, but they are literally taking their life in their hands every time they fly like that. The wake turbulence created by a B747 in a landing configuration at 180 knots could swat a Cessna 172 out of the sky like a newspaper swatting a fly.

If you learn nothing else from this book, learn to stay at least 5 miles outside of and 1500 feet below any arrival segment of a TCA. You can recoup the cost of the extra fuel that you will burn by cutting down on the coffee and the cigarettes. All three of these actions will increase the likelihood that you will bounce the grandchildren on your knee.

THE LONG WAIT TO CLIMB

Even though the above situations occur on a too-frequent basis, I think that there is one other circumstance that occurs around large airports that causes more frustration and pilot complaints than any other. While this is not directly related to the existence of a TCA, the traffic that is generated at TCA airports certainly exaggerates the problem. I am talking about the need to restrict altitudes and/or speeds of relatively high-performance aircraft operating at other than the primary airport.

These complaints occur because of the fact that most general aviation and business pilots operating out of satellite airports surrounding the primary airport do not understand the procedures that we use and why we use them.

Let's take a look at the Atlanta airspace configuration and discuss why we have to restrict the climbs or descents of aircraft operating out of our satellite airports. I don't mean to suggest that all aspects of the Atlanta airport operation or its TCA are typical of other airports, because they are certainly not. What is typical of our operation, however, is the fact that we, like almost every other ATC facility in the country, use structured routes into and out of our airports. These routes provide separation between aircraft even if total loss of communications and radar should occur. In this extremely unlikely event, most of the aircraft on an inbound or outbound route would be separated by nonradar altitude procedures until we could arrange alternative forms of control and communication.

One disadvantage of this type of route structuring is that the arrival and departure routes frequently have a steady stream of traffic along those courses. Aircraft wishing to climb or descend while transitioning across these flight paths often have to wait until they are clear of these routes before changing their altitudes in order to remain separated from that stream of traffic. Let's take a look at a case in point.

If you refer to the ATL radar map you can see that traffic inbound to the Atlanta airport is vectored along one of four flight paths depicted by the inbound arrows. This particular chart applies to landings to the east, so let's just deal with the traffic inbound for the base leg from the northwest. Atlanta accepts traffic from the ARTCC along this route at three different altitudes, 6000, 9000, and 13,000 feet MSL. Outside of the 25 DME arc (the dashed arc segment at the point of the arrow on the upper lefthand side of the chart) we vector aircraft, descend the higher aircraft to 9000 feet, and line them up in trail at the same speed. Once we cross that 25-mile line we establish the traffic on a generally south/southeast heading and establish the 6000-foot traffic in line with the higher aircraft which are now also descended to 6000 feet. The net result of this procedure is a steady stream of aircraft at or descending to 6000 feet and 5 miles in trail.

Now that you have the picture of this operation, let's take a BE90 aircraft off of Runway 8 at the Fulton County Airport (FTY), which is about 10 miles north/northwest of Atlanta (SEE DRAWING). Let's assume that this aircraft wants to go west and climb to 16,000 feet. If we turn that aircraft to the right towards the west, we will mix it up with the traffic turning onto an eastbound final approach course to Atlanta's northernmost runway. Turning in this direction means that the aircraft will have to be kept at a very low altitude until it is completely outside of the arrival course and it could also cause a real thrilling episode if the pilot were to make a wide sweeping turn as it goes to the west.

Consequently, the aircraft is turned to left to go west, and it can only be climbed to 5000 feet until it has cleared the arrival stream that we described earlier. Because of the traffic situation, this aircraft will have to remain level, at a low altitude, for 10 to 15 miles. This is a very inefficient operation for that aircraft.

If the weather is IFR and the pilot is in the clouds, he or she will never see the traffic. Additionally, all of the traffic lined up for the base leg into Atlanta

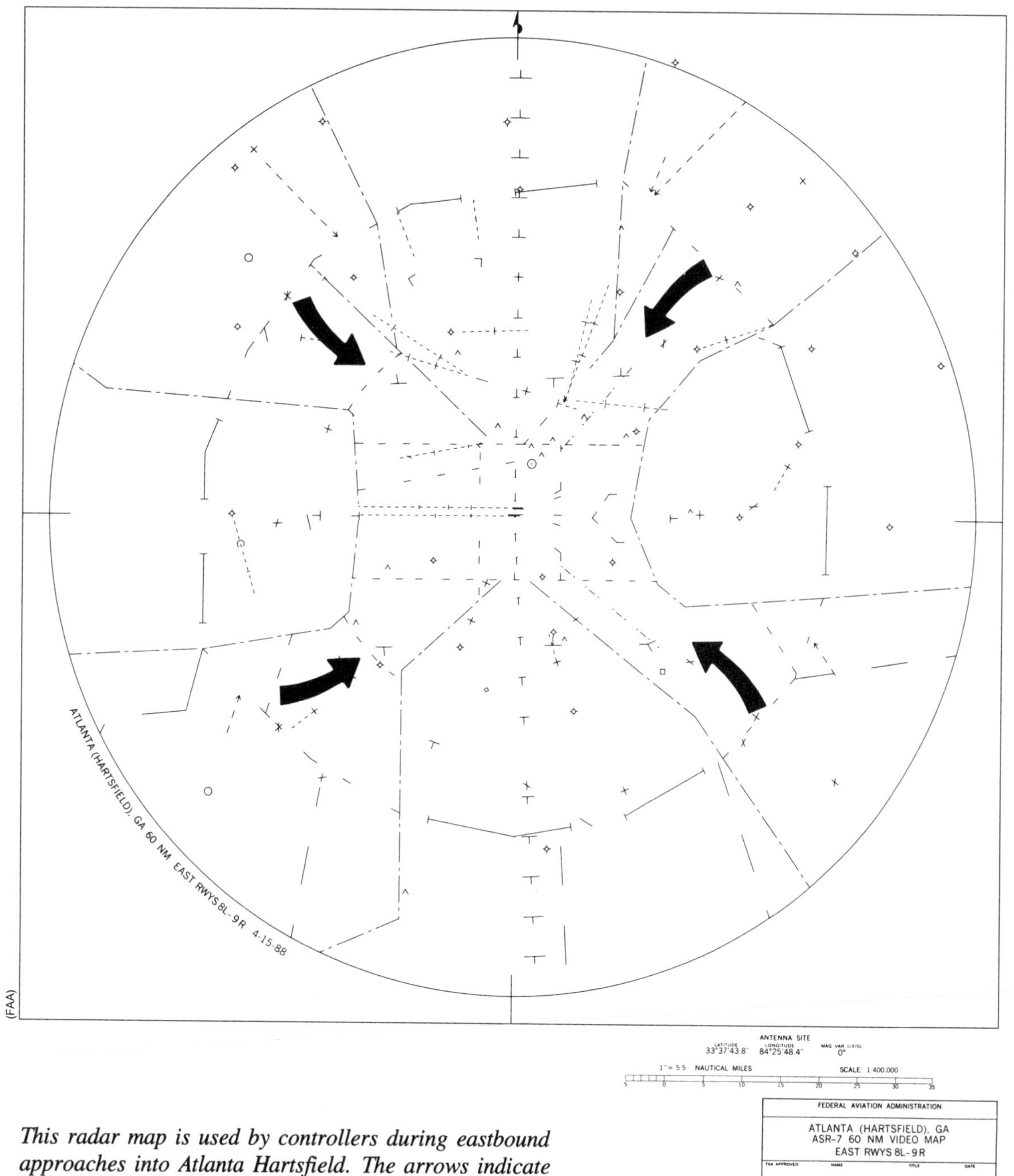

This radar map is used by controllers during eastbound approaches into Atlanta Hartsfield. The arrows indicate the four major arrival corridors.

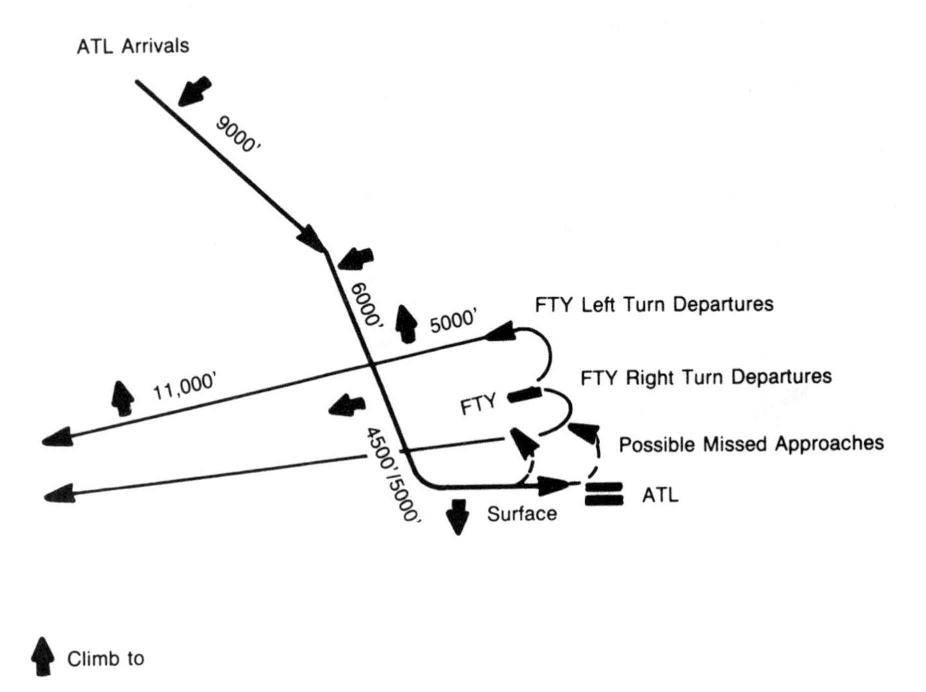

Westbound departures from Runway 8 at Fulton County (FTY) must normally make left turns and remain at 5000 feet or below until crossing the Atlanta (ATL) northwest arrival route. Right-turn departures from FTY would normally conflict with ATL missed approach courses.

is being worked by a different controller so the BE90 pilot will not hear them. If we assume that the controller working the BE90 is not very busy and is not making any other transmissions, you can understand when the pilot begins to wonder why he is being held down. If you don't hear it and you don't see it, there must not be anything there, right? This is where we start hearing transmissions like, "N5TA is *still* level at 5000 feet."

What makes this situation even worse is that this same pilot may have made the same exact trip earlier in the day when there was no inbound traffic on the base leg. In this instance, the controller would have coordinated a climb through the arrival controller's airspace, and the King Air would have made an unrestricted climb to the highest airspace that we own. If we did that this morning, then what we are doing now must be the result of lousy air traffic control or a controller who doesn't care about helping out the pilots, right?

The fact of the matter is that almost every controller, in almost every set of circumstances, usually goes out of his/her way to expedite the pilot unless there is a good reason not to. That reason could be conflicting traffic or a procedural requirement.

The situation I described is not an isolated case. Aircraft inbound to our satellite airports travelling through that same arrival corridor must come in at 5000, 7000 or 8000 until they clear the arrivals before we can change their altitude. Similar arrival and/or departure situations occur at almost every point in our airspace and in the airspace of every ATC facility in the country. In a TCA this type of situation is simply complicated by the volume of traffic, the complexity of the airspace, and the separation requirements of the operation. Often in TCAs, the controller may be willing but unable to accommodate the pilot's desires.

By simply radar identifying an aircraft or agreeing to vector an aircraft for a practice approach, a controller may be buying into a requirement for separation that current traffic conditions will not permit. We have to be very careful not to put ourselves in a situation where we buy off more than we can handle. As you will see in the next few paragraphs, sometimes just saying "stand by" can give some pilots the idea that they can do whatever they want.

"REMAIN CLEAR OF THE ARSA"

The *airport radar service area* (ARSA) is a form of controlled airspace that has been implemented gradually over the last ten years. Established around many moderately busy civilian and military airports, the typical ARSA is a two-tiered circle with the inner circle extending 5 nautical miles from the primary airport and extending upwards from the surface to 4000 feet AGL. The outer circle extends out to 10 nautical miles and upwards from 1200 AGL to 4000 AGL. Operations within these areas require that the aircraft be equipped with a two-way radio and that the pilot establish communications with the controlling agency prior to entering the ARSA and maintain communication once inside the ARSA.

ARSAs are an extension of the old terminal radar program of Stage I, II, and III advisories. This program was not mandatory on the part of the pilot, but it provided radar advisories and sequencing similar to the services provided to IFR aircraft. This program gradually evolved to the point where the limited services provided by Stage I were phased out and, with the exception of the establishment of a few TCA facilities around the country, terminal radar service areas (TRSAs) became the primary method of providing radar services to the VFR pilot.

As the complexity level and traffic requirements of certain airports increased, it was decided that an additional type of controlled airspace was required to handle the safety requirements around these facilities. ARSAs were established as a type of middle ground between the TCAs and the TRSAs. These new concepts in ATC mandated the services provided by the TRSAs without establishing all of the equipment and pilot requirements of TCAs. Some people say that these are actually the Group III TCAs that were never designated.

Beyond the grousing that some pilots like to do about the establishment of any type of controlled airspace, these airspace segments have generally been recognized as necessary or at least helpful in a busy air traffic environment. In general, the pilot participation and cooperation within these areas has been uniformly good. Unfortunately, there is always someone out there who likes to test the waters to see if they can walk on them.

The arrivals and overflights paragraph of FAR 91.88, which is the regulation that governs operations within an airport radar service area, states in part that "no person may operate an aircraft in an airport radar service area unless two-way radio communication is established with ATC prior to entering that area . . ." Let's take a look at the following ARSA communication sequence and see if the pilot has met that requirement:

Pilot: *XYZ Approach Control, this is N123TA, 20 miles east of the XYZ airport, inbound for landing with Information Alpha.*

Controller: N123TA, XYZ Approach, stand by please.

Pilot: *Roger, standing by.*

Has the pilot satisfied the requirements of FAR 91.88 and established communications with the controlling agency of the ARSA? The FAA recently decided that the answer is yes—that the pilot has met the letter-of-the-law requirements and that the above exchange constitutes an establishment of communications with ATC.

By using significantly different regulatory verbiage, the FAA has, intentionally or not, differentiated between the requirements for entering an ARSA and the requirements for entering a TCA.

The regulatory statement used in connection with TCAs states that a pilot must have "received an appropriate authorization from ATC prior to operation of that aircraft in that area." In order to impose the same requirement on an ARSA that is imposed on a TCA, it appears that the ARSA would need the same regulatory wording.

However, according to *FAA Aviation News*, if the controller simply replies, "Stand by," *without* saying the aircraft's call sign, the FAA feels that two-way communications have *not* been established and ARSA entry is not permitted.

COMMON SENSE

Our system of air traffic control is based on the concept that everyone will use some common sense while operating within the aviation environment. Frequent attempts to push the rules to their breaking points simply result in increasing the strength of the rules. I think I have shown you on several occasions that the FAA constructs its regulations in a way which intentionally allows as much flexibility as possible. This is done so that the controller and the pilot are not automatically in violation of a rule when they need to use a liberal interpretation of that rule in order to provide the services in a safe, orderly manner.

When people continually stretch these rules to the point where safety is compromised, the agency is left with no choice but to tighten the reins to the point where there is only one possible interpretation. This places a limitation on what we as controllers can do and just serves to fuel the arguments of those people who insist that we are trying to squeeze out the little guy. The agency really does try to allow for a little common sense when they build regulations. But, after being clubbed over the head in the courts and the press with the argument that the general public needs to be protected against those who lack that quality, the agency is forced to try to regulate common sense and good judgement. And that, my friends, is impossible.

If you listen closely to what a controller says these days you will notice that they frequently restate the obvious and double or triple verify that you understand what they are saying. Virtually every communication with a VFR aircraft by the Atlanta satellite control position these days ends with the statement, "Remain clear of the Atlanta TCA." This and other verification statements do not come about because we think pilots are stupid. Rather, they have come about because we, or others like us, have spent days in court watching pilots' lawyers use situations like the one mentioned above to circumvent the intent of a procedure. Most of us do not have a law degree, and we do not construct our control instructions with the idea that we may have to defend each one in court. But we do learn from what happens to other controllers who have had to do that, and we adjust our phraseology accordingly.

I have personally noticed that there seems to be a more adversarial, or at least a less trusting, relationship developing between pilots and controllers over the last few years. My contacts with friends on the other side of the radio indicate that they too are getting that same feeling. This is terribly unfortunate and one of the worst errors that any pilot or controller can make, because we truly need confidence and trust in each other to make the system work. I know I seem to be harping on this concept, but then that is the primary reason I have written this book.

We could probably pick out a few other airspace segments to talk about, but I think that by now you have begun to see the relationships among them and the importance of each. You may be one of those pilots who think that there are too many different types of airspace and too many regulations governing them. In some cases and to a certain extent, I happen to agree with you.

In fact, by the time you read these words, the terms TCA, ARSA, ATA, etc. may be obsolete. FAA is proposing to redesign and simplify the entire airspace system. If this proposal is adopted, you'll be flying in Class A, B, C, etc. airspace (like in Canada), airport traffic areas will become ''control tower areas,'' and I'll need to revise this book!

Regardless of these changes, we should all attempt to operate within the airspace system in accordance with the current rules and use them the way they were intended to be used. Doing it any other way limits the ability of others to do their job and will eventually exact its own price upon you.

7

The Future

SO FAR IN THIS BOOK I HAVE CONCENTRATED ON HIGHLIGHTING WHAT I HAVE called "pilot errors." As you have seen, these errors are, for the most part, ones of misunderstanding, mistakes, or lack of knowledge about a particular topic. For the last six chapters I have tried to relate a series of war stories or create conditions through which we can all see the traps and the subtle mistakes that can jump up to bite us. In this chapter, however, I will talk a little about the future of the FAA and of aviation in general and, as you read it, you might be wondering what that has to do with pilot errors. Try to look at this subject as a primer on preventive medicine, because today's lack of planning is tomorrow's error.

Perhaps the biggest mistake that virtually everyone in aviation circles has made, and continues to make, is that we lack either the vision or the ability to anticipate what will happen to our industry in the future. This failure ranges across the full spectrum of aviation circles and in some cases has been spectacular. We see it in the lack of planning on the part of airport planners who fail to obtain enough land for airport expansion or noise abatement. From there it runs all the way to the individual aircraft owner who decides to buy a big aircraft and equip it with cheap radios, rather than a slightly smaller bird equipped with a high quality, full-IFR panel.

We always seem to miss the mark when deciding what we can "afford" to do right now. As two old sayings go, "You get what you pay for," and "Pay me now, or pay me later." The future is the only area left where we can prevent mistakes, so let's at least try to understand a little bit about what's coming so we can make some intelligent decisions when faced with those changes. Now let's talk a little bit about tomorrow and let's start with the FAA.

FAA—CAUGHT IN A CRUNCH

Depending upon the person with whom you speak, the FAA is either at the forefront of futuristic aviation technology or is being dragged, kicking and screaming, into the 21st century. The actual truth of the matter is that, like any other aviation organization, we are probably somewhere in between.

When you ask the typical non-aviation-oriented person to describe what a control tower or an air traffic control center would look like, they frequently describe a star wars scenario with futuristic mockups and technology literally falling out all over the place. Surprisingly, a lot of pilots have this same impression of how it should look. Unfortunately, this will probably not be the case.

I remember watching the early space shots and the Skylab launch and thinking to myself that these vessels must be the epitome of modern hardware and instrumentation. Then, several years later when I finally got a chance to tour the equipment at the National Air and Space Museum, I remember thinking that it contained an awful lot of olive drab green and gun metal grey looking antiques that I probably wouldn't put in my Cessna. In my mind I had visualized this equipment as being futuristic, and as the years went by, I allowed my mental image to be upgraded with the changes in technology. Unfortunately, the actual equipment stayed the same. It was probably the state of the art in the 1960s, but this is no longer the 60s. I think that this is the typical impression that a lot of first-time visitors have when they come to an older FAA facility. After all, they pay a lot of taxes and expect to see state-of-the-art equipment everywhere. So, they don't really look close at what some of this equipment is capable of doing.

Don't get me wrong, a visit to a busy ATC facility is often an impressive experience to people who have never been to a place like this. These people are usually already in awe of the type of work being conducted there, and when they see the various banks of flashing lights, radar displays, and sophisticated-looking electronics, they usually leave wondering how mere humans could possibly master all of this knowledge. Those of us at the center of this adulation would, of course, not dream of bursting the impressions of these mere mortals. So we don't tell them that one of the flashing lights means that the telephone at the supervisor's desk is ringing.

Seriously though, the FAA is constantly trying to keep its existing components up to date and acquire new equipment before technology leaves us behind. Unfortunately, the process by which we currently plan for the future is itself a limiting factor in being able to deal with the future. Our budget process is one that asks us to envision what we think we will need three to five years in the future, estimate the extent of that need, and plan for what we think the cost of that equipment will be.

These plans are then incorporated into a comprehensive budget proposal and submitted to the Congress as part of the total budget process. Even if the budget is approved as submitted, we still have to adjust to changes in the acquisition process (over which the FAA has little or no control), and we have to react to sudden requirements to acquire new equipment.

If you can remember the drastic increases in interest rates and inflation in the late 1970s, you should also remember what that did to the prices of the things that you purchased. The same thing happens to equipment that we buy or order, and this is how a cost overrun starts to develop. Perhaps you can see where that leaves us when it comes to purchasing our supplies and equipment. When you are living on a fixed income and the cost of what you buy continues to go up, you have to make cuts someplace. The FAA's budget is similar to a fixed income and certain things, such as employees' salaries and the telephone bill, always have to be paid. When you begin to run out of money, the net result is that some projects are going to get delayed or canceled. Carry this logic out over a 10–20-year span and you end up with old equipment that requires massive changes to keep up with new requirements.

Now that you have an idea of how we get our funding and how that funding erodes over the natural course of events, add a little thing like Graham-Rudman to the picture. Your carefully planned budget has now been suddenly reduced by 10% after you have already committed every cent that you have. It doesn't stop there either. Congress sometimes gets into the act in even more subtle ways.

Imagine an organization with a budget of X dollars. This organization is currently spending 60% of its budget on salaries, and the rest is earmarked for every other part of its daily operating expenses and future needs. If the controlling agency for that organization votes its employees a 2% pay raise and then tells the organization to absorb the cost of the pay raise from its existing budget, something has to give. The 40% portion of the budget that was to go for equipment and supplies has suddenly been reduced by the cost of the pay raise. You can only rob Peter to pay Paul so many times. When Peter finally cries Uncle, one group in the FAA brain trust starts trying to figure out how to obtain a supplemental budget infusion, while another is trying to figure out how to cope with the fact that we will probably not get that extra funding. I don't envy the task or the job of anyone caught in this dilemma.

That's enough of the sad stories about the plight of the FAA. Let me now give you an idea of the different ways that the procurement process is accomplished and why it takes time to acquire new equipment.

Procurement

The FAA buys a lot of different things and we obtain them from a lot of different places. The generally accepted concept is that we announce that we want something, solicit bids for that product, and then purchase it from the lowest bidder. This is a true statement, but it is not entirely accurate. Some of the things that we use are covered by a patent or a copyright. If you want that product, you must buy it from the organization that holds the paper on the device. Additionally, there are not a lot of telephone companies, electrical power companies, and other utilities from which to choose when you need those kinds of services. So you must deal with that reality when you plan those services. Congress sometimes mandates that we will obtain particular pieces of equipment, and these types of components are frequently manufactured by only a few organizations. In this case you simply do what you are told.

In general though, we do go through some kind of bid process to obtain what we use. I had the pleasure of serving on one of the design teams that developed the specifications for a component called an *information display system* (IDS) that is currently being installed in some of the facilities in the FAA's southern region. I got a chance to see firsthand one of the processes for designing a piece of equipment and selecting the winning bidder for that component.

The process begins when a need is identified for a piece of equipment that will provide a particular service. In this case, controllers have a need for ready access to a large amount of information, such as weather at several locations, instrument approach profiles at 10–25 different airports, and a list of the frequencies for several different navaids that are scattered over several hundred square miles. They also want to be able to obtain this information without having to search for the data, move to several locations to obtain the data, or thumb through a large document to find the appropriate details. This information needs to be placed in a convenient location, in a format that is easily readable, and it must be able to be changed as quickly as the changes occur. If you consider how often the weather changes at five or six separate locations, I think you can understand the importance of this feature.

Once a decision is made that we need to develop a solution to meet this need, a group of engineers studies the problem. They interview several controllers concerning their desires, survey what technology is currently in use, and design a draft component that solves the problem. They lay out the basic technical specifications that are required and add to these a group of preferred features that could either be incorporated into the specifications or be available as modular additions.

These specs are then reviewed by a group of people whose areas of expertise cover the fields of procurement, funding, research and development, and controller use of the equipment. This group takes the original specifications and suggests any changes to the design that would make it as cost effective and functional as possible. The engineers then review the specifications against the "wish lists" that we provide, and develop a final set of design specifications that are submitted for bid to interested companies. These design specifications contain detailed requirements for everything from the size and type of parts that will be used in the component, to the level of quality control that is required from the company that receives the bid. The process then takes one of two different paths.

If such a component is already on the market or is adaptable from one that is, an individual company will submit its bid based on the adaptation of that equipment to our specifications. This bid usually includes the cost of the individual components; the installation, training, and maintenance costs associated with the original purchase; and either a long-term maintenance agreement or a proposal for training the FAA technicians who will maintain the equipment.

If the component does not currently exist, interested companies may decide to compete for the contract bid by designing the system and building mockups that demonstrate the qualities listed in the specifications. In this case, the specifications are usually somewhat more general and they allow the company enough flexibility to use existing systems combined with new technology to go beyond the minimum design specifications. The proposed bid, which is similar to the one described earlier, is then submitted to the FAA for consideration.

The FAA review committee then selects the design which: (a) meets the minimum design specifications and (b) is the lowest bid from among those meeting the specifications.

This process takes time and, depending on the sophistication of the individual component and the need to develop new technology, it could easily run into several years before the actual equipment appears in the ATC system. Just to complicate the process, let's add a couple of other facts. There are literally hundreds of these development projects, some of which get canceled for various reasons. Other components have to be redesigned when major breakthroughs are achieved in their specific technology or they are made obsolete by other equipment. Finally, all of them have different time frames for installation, which means that we sometimes have to bridge the gap between the installation of related equipment by upgrading the portion of the existing equipment that has to serve longer.

As a result, the typical FAA facility is a mixture of newly designed, up-to-the-minute equipment and older but still serviceable components. This older equipment has usually been modified to the point that, occasionally, it barely resembles the original design. Some of these components are no longer being built by the manufacturer or, in some cases, the manufacturer is no longer in business. It is only the abilities of a superlative staff of technicians that keep the equipment going while maintaining one of the best on-line service records of any industry.

We have some people who work for us who have done a magnificent job of maintaining equipment and others who produce the same level of work in planning for our future, but I think that even they would tell you that this procedure is a problem.

Just think back to the early 1980s and take a look at where the computer industry was at that time. Aside from a few geniuses, I doubt that anyone could possibly have guessed the quantum leaps in technology that occurred since then. Anything that we could have ordered or purchased back then would be a child's toy in today's market. In fact, the very computer system that I am using to write this book would have been something that existed only in the minds of computer theorists at that point. Much the same can be said for materials used in navigation equipment, communication capabilities, and some of the materials that make up the aircraft that you fly.

In a few years we may be designing aircraft made of materials that do not exist today. We may be using energy from storage cells and photovoltaic systems made of ceramic superconductor material that can generate and hold more energy than we could possibly need. We may even be linked to all of these systems through equipment that transmits brain waves to mechanical servos. The possibilities are only limited by the barriers we place in our own minds.

The point I am trying to make is that telling you where we will be in the future is as difficult as guessing the roll of the dice. The future is an ongoing operation in an organization the size of the FAA, and the very process of replacing outdated equipment takes us in that direction. An example of this process, and something about which you may have heard, is the new IBM *host computer system* recently installed throughout the nation. These new mainframe computers, like the 10–15-year-old ones they replaced, connect each ARTCC to the terminal facilities within its area. (Remember what I said about computer technology over the last few years and you will know how much progress has been made during that time.)

The host computers were in development and testing for at least 10 years. This process might appear too long and drawn out, but the host computers have to interface with several other types of computer components already in use, and they have to be compatible with other components on order. For example, they have to work with the ARTS and the FDIO systems. They have to interface with existing and planned FSS input terminals, ARTCC sector computer input devices, and the new sector suite equipment that we will discuss later in this chapter. Most importantly though, they have to work the first time they are turned on and every time after that.

TAKING THE INDEPENDENCE AVENUE

One of the most important possible changes in the FAA is only indirectly related to its equipment. There is currently a move afoot within Congress that has strong support from some powerful segments of the aviation industry to take the FAA out of its current status as a sub-cabinet-level government agency and make it a completely separate organization. If it occurs, it would be the biggest change to the structure of the FAA since its inception. Suggestions range from making it a full cabinet-level organization, to making it a private corporation funded by user groups.

If you keep your ear to the pulse of the talk on this subject, which I assure you *we* do, you can hear several ideas being kicked around. One suggestion would change our organization to a semi-private corporation similar to the U.S. Postal Service. Others suggest that we become a completely privatized organization run by a management group answerable to user groups. The proponents of this idea believe that we would then be better able to respond to the pressures and changes of the aviation market, to fund our requirements, and react to changes in technology.

The trend of thought in these proposals seems to focus around the idea that if we have the ability to buy equipment off the shelf without waiting five years for it to become obsolete, we will have solved all of our problems.

Here again, the real answer is probably somewhere in the middle. An unlimited budget would certainly help reduce the time required to update equipment, but that is only half of the picture. Even without the absolutely newest equipment, the agency has done a fairly good job of keeping up with the demands of the traffic system. Let's face it, ATC all over the world is based on our model. Many of the limiting factors that currently bog down the system have nothing to do with the actual control of air traffic, and some of these factors will not be affected by changes in FAA funding or management.

For example, weather can limit the number of operations that can be conducted at any facility. When an air traffic operation is hampered by thunderstorms or forced to reduce operations as a result of snow and ice, arrival and departure rates go down. Another factor is scheduling. Most Americans want to arrive at their destination with enough time left in the day to do their work or return shortly after the workday is over. The result is a series of heavy rushes followed by relative lulls in the traffic picture. A third factor is simply the lack of places to land or depart that are within a reasonable distance from where we want to be.

As the "superfacilities" demonstrate on a daily basis, only so many aircraft can land or depart in a given period of time on a given number of runways. But just doubling the number of runways at a place like Atlanta will not automatically double its capacity. You also have to figure out a way for the arrival and departure airspace to accommodate the traffic.

We are, of course, always fine-tuning our procedures to squeeze out a little extra production. We also try to teach our newer people how to run a consistent interval and take advantage of every technique for the best operation. If we deal strictly with the operational aspects of increasing volume, the future lies in refining our separation processes and standardizing our procedures. Think back to the earliest days of air traffic control and look at the separation standards that were used. We separated aircraft by counties and states compared to the miles and feet that we now use. Examine the trend and I think you can see that as technology improves, more aircraft can be fit into the system.

Given the extra runway capacity and some new procedures and technology, we should be able to figure out a way to increase that productivity. Unless we can learn to control the weather, change the travel philosophy of the American public, build facilities where they are needed, and significantly redesign the technology to allow a reduced separation interval between aircraft without reducing safety, we are going to continue to operate on the ragged edge for some time to come.

The FAA has also done a reasonably good job of spreading the talent of its people where it does the most good. The techniques learned at the larger facilities gradually filter down to other facilities as we assign some of our better people to be the supervisors and managers of those facilities. These smaller facilities are usually the starting points for some of our new controllers, and while they are brimming over with youth and enthusiasm, they lack the experience of the older controllers. Those who have been down a road before can teach from experience and by example. Hopefully, the new people will learn some of the old dogs' tricks.

This knowledge and ability flow is by no means a one-way street. Sometimes the old dogs learn a few things themselves. Some of the best ideas and techniques that are eventually implemented at all facility levels come from the minds and the experiences of the people that we have selected from smaller places. Those of us who have been at a large facility for a long time sometimes cannot see the forest for the trees, and new people come up with obvious answers and fresh ideas.

Restructuring of the FAA has a lot of potential and is gaining some support from very powerful elements. Let's hope that they keep what is good about the current FAA, design an organization that deals with the human needs of its employees, and structure the agency to be able to respond quickly to change.

CONTRACT EMPLOYEES

Along with the philosophy of taking the FAA out of the government and into the private sector, another idea that is gaining some momentum in the air traffic control business is the concept of contracting out a lot of the work that we currently do with agency employees. While this causes some concern among those of us who still work for Uncle Sam, I can see the financial advantages which bring about this trend. A major cost of employing an individual comes from the expenses associated with retirement plans, health and life insurance, and the cost of administering such a large number of employees. Consequently, a 100-person reduction in the agency's work force saves considerably more than the salary of 100 employees. If, for example, it costs the agency an additional 30 percent of an employee's salary in the form of benefits, but the same job can be contracted for that salary plus 20 percent, isn't the choice a matter of simple economics?

Currently, much of the day-to-day work associated with the operation of computer systems at some ARTCCs is being handled by contract employees. Several of the Level 1 (the smallest) control towers throughout the country are staffed by non-FAA employees and operating under a contract. There are also several ongoing studies concerning additional contract possibilities that include the ATC training programs, some maintenance capabilities, and additional low-level control tower facilities. It is estimated that, by mid-1990, 30 Level 1 towers will be operating under a private sector contract, with the remaining 100+ facilities being examined for similar changes.

One of the prime disadvantages of this contract concept is that it also eliminates the existence of the "company man." I don't think that a hired gun ever brings the same dedication to a job as someone who has nurtured a project through from the beginning does. Contrary to a lot of public opinion, most FAA employees usually give about 45–50 hours of work for every 40 hours of pay. They do a lot of work and thinking at home, and the very important phone call always seems to come at quitting time. While I understand the logic behind contract workers, I hope the FAA continues to keep its employees and, for that matter, the entire aviation industry informed of the actions that it is taking. These actions affect everyone's planning for the future and they do tend to shake the security blankets that we all carry.

RADAR ADVANCES

Let's get back to future technology and talk first about radar. When I use the term future, as it relates to radar, I am not necessarily thinking about things that are still in the theoretical or development phase. The new ASR-9 radar equipment for terminal radar sites and the new ARSR-4 long-range radar equipment for ARTCCs are good examples. These two components of the next generation of radar equipment already exist. The "future" in this context is the date that they, or other similar systems, will be installed nationwide. Remember, you just can't take a piece of radar equipment out to a site, slap it together, throw a switch, and call it operational.

The ASR-9 and ARSR-4 are typical of new ATC equipment in that they have the following design specifications:

- They have solid state construction.
- They use remote maintenance and repair techniques to some degree.
- They are of modular construction, allowing expansion and growth.
- They can be tailored to site specifications, allowing them to be built at almost any location.
- They are supposed to drastically improve radar coverage and detection.

Sounds like a commercial doesn't it? Maybe, but these are some of the specifications that the FAA has in mind when they ask a company to design a system and submit a bid for equipment that we will use into the next century.

Of all of these wonderful specifications, the one that appeals to me the most is the one about site environments. If you remember our discussions about the problems with radar, you will remember that obstructions to the line-of-sight signal cause us the most problems. If we could build our radar sites on top of high hills or tall support structures, we could eliminate the majority of those problems. Of course this would require a little bit of a look-down capability in the equipment and we might begin to see things that would scare us, but I think we can handle the technical part and I'm willing to take the chance on the other.

I think the newer generations of surveillance radar hold one of the keys to tomorrow's air traffic control. Part of the antenna array on these systems is the beacon radar system antenna. This is the part that interrogates the transponders, and the newer ones are compatible with the new Mode S data-link transponder. When fully operational, this equipment will allow direct communication between landside and compatible airborne computer systems. We will then be able to send and receive data and instructions via keyboard consoles, which will gradually eliminate some of the frequency congestion that currently slows us down. Once this air/ground data link is established and proven, our ability to communicate with the pilot and/or hardware that is flying the aircraft will be greatly enhanced.

Eventually this equipment and other data-link devices will connect inertial navigation equipment on the aircraft to ground-based guidance systems for navigational assistance. I can also foresee a system which, when linked with satellite relays, will allow communication between aircraft on the ground and controllers miles away. We will eventually conduct long-range transfer of data associated with changing weather and runway configurations. Devices such as ATIS broadcasts could be received and verified without any audio communication between the pilot and controller. Both pilot and controller would be ready to conduct some serious air traffic control the second that they started talking to each other.

For now, this system will be limited to transferring control instructions, such as altitude assignments, and it will be tested with verbal confirmation until everyone is satisfied that it works 100 percent of the time. But the capability of this system for communications of all kinds is unlimited, and it will eventually be used to its full potential. When that happens, air traffic control will not even remotely resemble what we do today.

Super Scopes

The rate of technological change is not limited to what we will see *on* our radar displays. It is also going to make a major difference in the surveillance radar display equipment itself. The radar unit, or perhaps the correct term should be video display unit, and the depictions that I will see on the scope are going to look more like my visions of a star wars console. The equipment will provide me with full-color presentations, enhanced graphics, digital displays, and selectable functions. The design of the equipment is aesthetically pleasing and very impressive, but some of the biggest changes are going to be on the inside.

If you look at the radar scopes of tomorrow and compare them to the ones used today, you will still see a video display unit that has many of the same external features. About the only thing that will set them apart on the surface are the newer switches, selectors, and hi-tech materials. But the information that will be presented on these new displays can easily be compared against existing capabilities by remembering what has happened to video game technology in the last 15 years. Remember the old ping-pong games of the 70s? If you go down to the video arcade today, you will see operator interactive games that change completely with every 50 cents that you put in the slot. Most of this change has come about because of the improvements in the graphics capabilities and the light transmission media that are being used inside those screens.

The theory behind the two generations of this equipment remains the same, but the application is like comparing TV sets of the 1950s to the ones we use today. Patterns of illuminated scanning lines called *rasters* were used to display the moving ball in the old ping-pong game. These rasters are made up of points of light called *pixels* which, in very simple terms, are single points on any video screen that can be illuminated on command. These little dots are then combined to form a picture, a number, or a letter when the proper sequence of pixels was illuminated. Sounds simple doesn't it? The problem was that the older radar systems could only address blocks of pixels instead of the individual points. The resulting display was not sharp and the movement was jerky. The screen definition and graphics capability resembled what you remember from the old pong game.

The improvements in raster technology since those days now allow each pixel to be addressed individually. These millions of little dots of light can be illuminated in various colors, in combinations that allow sophisticated graphics, and with a speed that is almost beyond the capacity of the human mind to grasp. The changes aren't just limited to what you see on the screen. The screen itself has been drastically improved. The old vacuum tube display systems have been replaced by advances such as plasma displays, which contain an ionized gaseous medium (composed of electrons and positive ions) that is essentially electronically neutral. This almost eliminates electronic noise and random static interference, and the resultant presentation is so sharp that you can almost feel the difference.

These and other changes have made an impact in almost every aspect of the video display unit. In terms of commercial application we are beginning to see 1- or 2-inch-thick television screens that resemble wall paintings. We have television sets now that will fit into your shirt pocket. Regardless of the application, be it TV or radar, it can be smaller, more powerful, more defined, present a sharper image, and allow the full use of developments in raster technology. All of this new technology is coming together to provide the controller, and in some cases the pilot, with a real-time, understandable view of what is going on.

ASDE

Another type of radar system that has been going through a major overhaul is the *airport surface detection equipment* (ASDE) radar. This piece of equipment is the one that allows controllers in the tower to "see" aircraft when the visibility is reduced to the point that you can't see the railing around the tower cab. The current system is housed in a round dome on the top of the control tower and it depicts a fuzzy outline of the principal features on the surface of the airport. At best, it shows the movement of aircraft and vehicles in a manner similar to the primary target on an ASR display. We do not have the ability to positively identify these targets, and there is no capability to place an alphanumeric tag on them. Plus, the entire system is subject to interference from moderate rain.

The newer ASDE system looks like the dish antenna on the top of hurricane-hunter aircraft (SEE PHOTO). The interior and surfaces of the dish are heated, it has a window through which the radar signal is broadcast, and it rotates to help sling off ice, snow, and water. As it is currently being installed, the ASDE 3 will still not have the ability to individually tag aircraft with an alphanumeric identifier. This is not because of a design deficiency in the equipment, however. Rather, it is because the system is a look-down type of radar, and the transponders on most aircraft are under the wings or the fuselage and would not "see" the interrogation signal.

It could be modified to track individual targets if someone were to specifically identify which target belonged to which aircraft, but this defeats the purpose of the system and is extremely wasteful of manpower. Once the new Mode S data-link transponders come on line however, this system will have the capability of automatically tagging targets with alphanumeric data. Finally, this system uses a technique of varying the frequency of the radar signal at regular intervals which makes it much less subject to the white-out effect of heavy rain. For facilities that frequently experience periods of reduced visibility due to fog or snow, this component will greatly enhance our ability to move aircraft. It will mean the difference between operating by pilot reports, similar to the pre-radar days, and moving the traffic as if the visibility is 15 miles.

SATELLITES

A great deal of information transfer is currently being conducted through the use of satellite relay, and some of the technology is beginning to find its way into aviation usage. Perhaps the most immediate use for satellites will be in oceanic air traffic control. A concept called *automatic dependent surveillance* (ADS) will

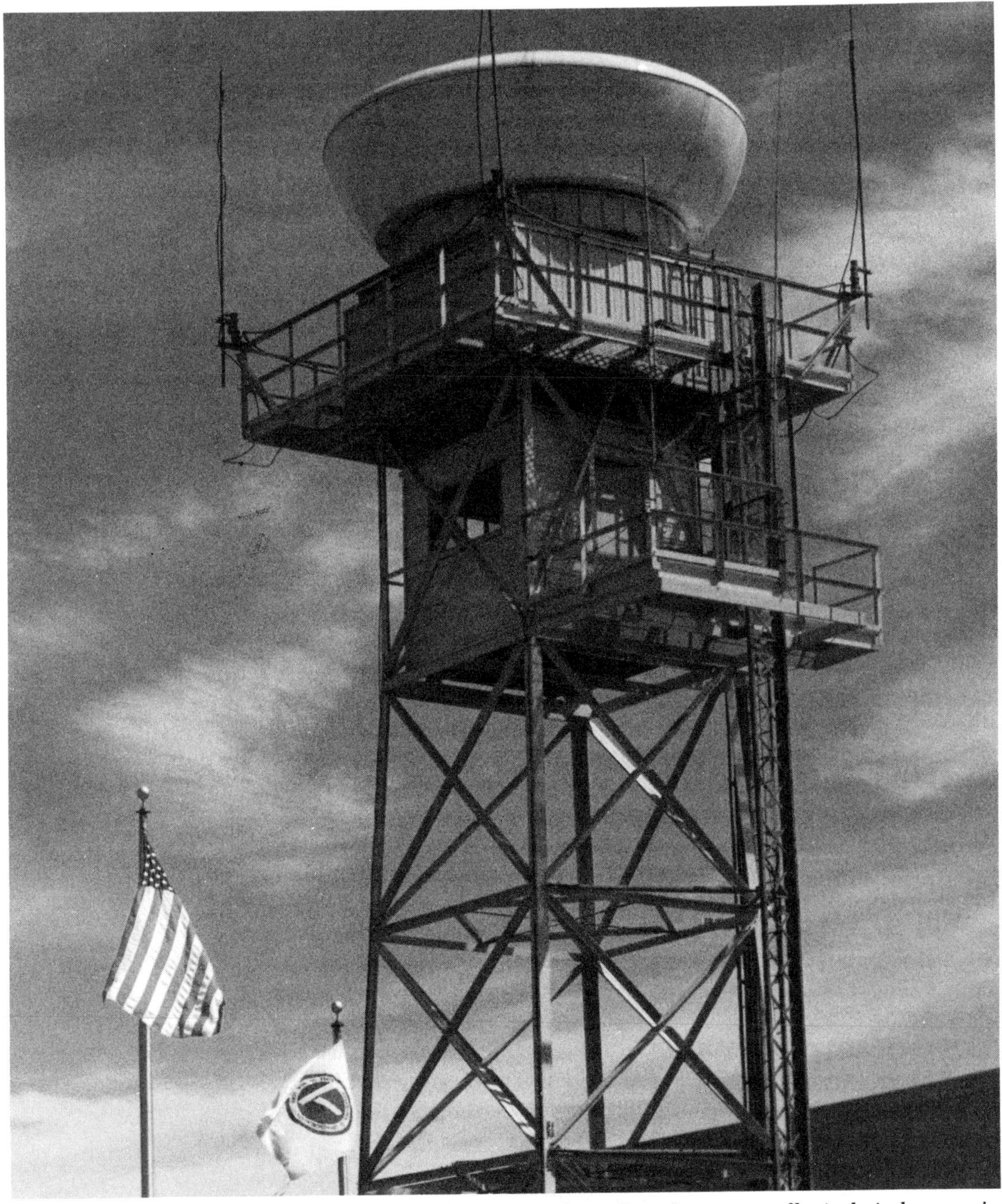

Advanced airport surface detection equipment (ASDE) radar will perform more effectively in heavy rain than existing equipment does. Eventually ASDE will be able to tag targets with alphanumeric data.

place aircraft in direct communication with a satellite throughout their transoceanic flight. This data-link communication between aircraft, satellite, and ground station would allow ATC oceanic sectors to have a computer-processed track on that aircraft, similar to a radar track.

One of the projects under study is called the *oceanic display and planning system* (ODAPS). Once tested and proven for accuracy, this system, or one very similar in design, would allow a significantly reduced separation standard to be used in oceanic ATC. We currently use separation standards of 60–100 miles between aircraft over the ocean. If we could cut that requirement in half, we would effectively double the system capacity. It is not outside of the realm of possibility that we could go well beyond that standard, perhaps even down to the 5–10 miles that we currently use in ARTCC sectors.

We are also beginning to see international cooperation with the use of satellite equipment, and this development literally opens up a whole world (no pun intended) of possibilities. Perhaps the best example of this cooperation comes from a series of incidents where the Soviet government provided information to our government concerning the location of emergency locator transmitter (ELT) signals that their satellites were receiving. These actions helped find several lost aircraft and probably saved several lives.

Several inertial navigation systems currently in use (mostly in the military at this time) must lock onto a satellite to fix their location. Once these data locks have been accomplished, the computers are capable of flying the aircraft only a few hundred feet above the ground at near-supersonic speeds. While this might thrill the adventurous soul, the average airline will probably use this technology at a more sedate altitude and a slightly more leisurely speed. But use it they will.

Eventually, the use of geosynchronous satellites (satellites whose orbit allows them to remain stationary over a particular point on the surface of the earth) with this technology will lead to the development of programmable, plug-in computer modules for any possible route of flight. These *course cards*, which would be downloaded from a worldwide master computer system, could then be plugged into an aircraft inertial navigation computer. They would contain the most up-to-date information about routes, navaid outages, weather, and inertial guidance information for virtually any route of flight. They could also be linked to a satellite update frequency which would incorporate changes as they occur and make adjustments as necessary. Pilots could obtain them at airline scheduling offices, FSS facilities, or perhaps over the phone from businesses specifically established to provide such a service. Again, the potential use of this equipment is limitless. If they can just figure out how to have it brew a good cup of coffee, it would sell like hotcakes.

TCAS

The changes that are scheduled to occur during just the next few years are going to drastically alter the way that pilots and controllers are going to do their jobs. Those of us who grew up in the needle, ball, and airspeed days had better learn to adapt to these changes or we will be reduced to the status of observers.

In today's ATC system, the pilot responsibility for separation relies strictly on the old "see and avoid" concept. Unfortunately, with increased onboard equipment requirements and the tendency to design airliners with a two-pilot cockpit, less time is available for pilots to look out the window for traffic. Part of the solution to that dilemma is called the *traffic alert and collision avoidance system* (TCAS). As currently designed, this is an airborne collision-avoidance system which relies on the reception of radar beacon signals from other aircraft to warn the pilot of conflicting traffic.

The first of these systems, TCAS-1, only generated traffic advisories, and the pilot was required to analyze the situation and take corrective action. The newer TCAS-2 actually offers the pilot conflict resolution in the form of verbalized maneuver suggestions which tell the pilot the best course of action for avoiding the other aircraft. These suggestions are similar to the "pull up" commands associated with a low-altitude situation on today's flight director systems. The alarms and the verbal suggestion information on this equipment are real attention getters for the pilot and significantly add to the safety of the flight. Even newer systems such as TCAS-3, which provides conflict resolution on all axes of flight, are on the drawing boards. I can envision the day when all aircraft are equipped with even more-advanced TCAS components. By that time, the computers on board the aircraft will communicate with each other, agree on a conflict resolution, inform the pilots of the situation, and adjust the flight path of each aircraft to compensate for the conflict.

TCAS requirements will take effect for airliners and commuters in the 1992–1995 timeframe.

MLS

Another of the more interesting concepts in the next generation of aviation equipment is what might happen to instrument landing systems. One example of this new technology, called the *microwave landing system* (MLS), offers some rather unique flexibility over the old ILS procedures that could significantly alter how we conduct instrument approaches.

If you visualize an ILS final approach course, you have a straight, narrow corridor extending up from, and along the extended centerline of, the runway. This corridor gradually increases in altitude (decreases if you are flying toward the runway), beginning at the glide slope transmitter and extending upward at a very shallow angle (2–3 degrees) for about 10 miles on final. The design of

Above: Existing instrument landing system (ILS) equipment limits aircraft approaches to one narrow, straight-in corridor.

Below: Proposed microwave landing system (MLS) equipment will permit an infinite number of approach choices, including curved approaches.

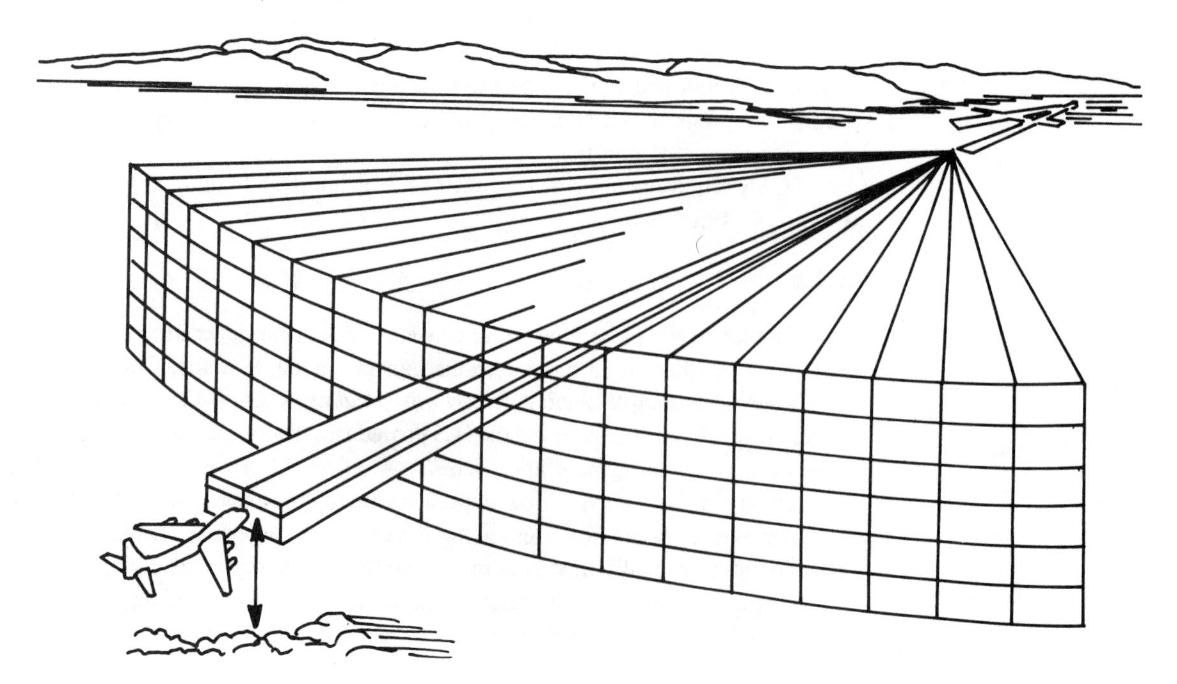

this equipment not only limits how we conduct our instrument approaches, it also effectively limits where we can build our runways. Remember, we have to, or at least we're supposed to, plan for the land required for instrument landing systems associated with a runway, when we plan the runway. MLS will change how that planning is done.

MLS will allow us to take this perfectly straight localizer/glide slope corridor and bend it into a curved path. Picture a corridor that starts at the runway, goes out along the extended final for three miles or so, and then gradually curves along a standard-rate turn until the entry point is on the downwind or the base leg. Controllers could vector an aircraft onto the downwind, have them intercept the MLS signal, and know exactly what flight path the aircraft would follow to the runway. A good flight instructor could practice instrument approaches and touch-and-go landings at the same time. But the most important aspect of this equipment is that, when the equipment is fully operational both on the ground and in the aircraft, multiple flight paths will be available to both the pilot and controller. We can vector to intercept an MLS approach to one runway from several different directions at the same time and each of these flight paths might have a different descent angle. Air traffic control has been equated to a three-dimensional chess game where all the pieces move all the time. This equipment will only reinforce that image.

We could build a runway just a few miles from a mountain and construct an MLS installation that will work for that configuration. The MLS flight path on the side of the airport closest to the mountain would allow you to fly a curved path along the foot hills of the mountain. For STOL aircraft, a separate "dive-bomber" course over that mountain could be designed. The number of potential flight paths is only limited by the volume of the traffic into the system and the sophistication level of the MLS unit on board the aircraft. MLS ground equipment is designed to be installed on a free-standing tower with a solid-state, remote maintenance, fully redundant electronic design that checks its own accuracy and notifies a beeper-equipped technician that service is required. Think of the applications that this equipment might have in Alaska alone.

MLS systems may eventually replace the existing ILS network, but their best theoretical short-term application seems to be in opening up some of the airport capacity that is currently not being used. Several of the busiest airports in the country have runways that are not now equipped with instrument approaches. Others have nonprecision approaches that become virtually useless due to frequent low weather. MLS systems added to these runways would increase the acceptance rates at each of these airports and enlarge the overall system capacity.

Another use for MLS components that could increase capacity even more, would be the establishment of a network of reliever airports that have the capability of landing aircraft in poor weather conditions. A large number of these small airports are relatively close to the major terminals and business centers in this country, but currently have no instrumentation for poor weather operation. For significantly less than the cost of building a new airport or runway, they could be instrumented and used as reliever airports.

Many business executives who fly their aircraft into the big city airports would rather go to the smaller airports, because they are often closer to their businesses and offer a more rapid turnaround for their trips. In this day and age of expanding the system's capacity, this concept is beginning to get a lot of press and the designs are starting to show up in regional development plans. The next time you consider buying some ILS equipment for your aircraft, check the market to see if MLS equipment has been developed to the point that it is beginning to be installed at a large number of facilities. If this component is available and beginning to replace ILS, the extra expense of an ILS/MLS component might prove worthwhile in the long run.

WEATHER TECHNOLOGY

In the chapter on ATC equipment, I explained some of the problems that we encounter with weather and how it affects our operation. Since it is not likely that we will be able to control the weather anytime soon, the next most logical step is to be able to see and forecast the weather and disseminate those observations and forecasts quickly.

The FAA is making significant strides in this direction. Several projects, either under development and installation or still in the research phase, when tied together will provide a broad range of weather services for the pilot and controller. The one that seems to get the most press is *Doppler radar*, so let's talk a little bit about that one first.

Most people think of Doppler radar strictly in the context of wind shear prediction, but this is a little misleading. First, let's talk about what Doppler actually is. The *Doppler effect* is described as an apparent shift in the frequency of a wave, varying with the relative velocity of the source and the observer. These energy waves can be generated by sound, light, or radio. The motion of water droplets as they move through the air, or the actual molecular motion of the air itself, such as in a wind gust situation, can be measured by observing such frequency shifts. The new weather radar systems under development use this Doppler measurement concept as a part of their information processing technique, hence the term Doppler radar.

The systems that are under development now are appropriately generically called *next-generation weather radar* (NEXRAD). NEXRAD components will use the latest technology in computers, video display, and graphic enhancement to present the best possible picture of the weather systems. One segment of NEXRAD, the *terminal doppler weather radar* (TDWR), is the component that is primarily designed to deal with wind shear events. TDWR systems will be installed in control towers across the country, and this is the radar system that the public will probably identify as Doppler radar.

Most of the hardware for this system focuses on the measurement of wind fields in the terminal areas. Its primary functions have already been demonstrated, and equipment is currently being tested. Additionally, considerable work is being done on the certification of onboard wind shear detection devices and ground-based sensor systems for those sites where TDWR is not feasible or cost effective.

Earlier I alluded to the fact that Doppler radar has developed an image of being able to predict weather events. It will provide several major advantages and improvements for controllers, but it will not directly *forecast* weather. Doppler radar is exactly the same as any other radar in the sense that it sees what is actually happening. It cannot predict or forecast weather and it is certainly not the cure-all to weather-related problems. It is still a measuring device and the ability to forecast with this system continues to rely on the expertise of the individual using the components. It does have an advantage in that it is significantly stronger and it provides the forecaster with much more exact information about air motion and precipitation. It also provides this information over a wider range of frequencies. Given these factors, good computers, and a wider network of more up-to-date information, forecasters can increase not only their accuracy but their speed as well. The contract award for the first of the Doppler radar components should be made late in fiscal year 1989 or early 1990 and the equipment should start coming on line in the early 1990s.

The development of new radar equipment and other weather associated technology does not just stop at the development of NEXRAD. A weather radar support facility was recently established in Huntsville, Alabama, to continue the work on evaluation of radar performance. The work that will be done at this facility will deal with theoretical concepts such as weather algorithms and several other things that are way above my head, but I think that you can see that much more is yet to be discovered where radar is concerned. There are also several certified crazies, operating out of the Oklahoma area, who chase after tornadoes to see if they can figure out how to predict these devastating weather events. These and other folks who work for the National Weather Service (NWS) have my respect and admiration, and I am impressed with what they are doing.

Some of their work is beginning to show up in the form of new equipment coming off the drawing boards. The NWS/FAA is beginning to contract for the installation of several components of the *automated weather observing system* (AWOS). When associated with the National Airspace Plan, these components will permit the collection of data from remote sites. These sites will then process the data and use computer-generated voices to transmit the information to pilots. Eventually, all of these components will tie together to form a nationwide weather information network. The long-range concept of weather reporting will connect the Doppler components of NEXRAD, AWOS, and local weather radar units together to a central weather processing (CWP) system through a series of meteorological weather processor (MWP) and radar weather processor (RWP) components, delivering information to where it will do some good.

This network will connect to a weather communications processor/data-link system that will tie to the air traffic control capabilities through Mode S transponders and the projected air traffic advanced automation system (AAS). We will talk about the AAS in greater detail later in this chapter, but I think you can begin to see the potential for both the pilot and controller to obtain more realistic weather data through this national weather network. Looking at it strictly from my perspective, it's really exciting. TDWR will provide me with the ability to virtually "see" wind shears develop, and the combined system will provide me with a weather mosaic through my new control position capabilities. If I can look west out of Atlanta at some weather 40 miles away and then look east out of Birmingham at the same system, I will be able to know exactly what the real picture looks like.

SECTOR SUITES

How, you might ask, is the controller going to be able to use all of this equipment? After all, he or she is only one person and what I have talked about so far is an awful lot of hardware. Well, we talked about the reduction in the size of the equipment and this is probably the second most important change to this equipment that has occurred over the past few years. These changes have made possible the air traffic control position of tomorrow.

In the future—which, by the way, is only a few years away—the controller will have a position of operation called a *sector suite*. This position is about the size of a large desk and it will contain equipment that could not have fit into a warehouse a few years ago. There were several different competing designs for the sector suite equipment, and IBM won the contract.

The sector suite concept is actually a grouping together of several video display terminals and keyboards around a controller so that he or she has immediate access to multiple functions. One builder's design has three terminals arranged in a semicircle around a controller so that all the controller has to do is move slightly to be facing the individual components and their associated keyboards.

Each major terminal has a primary function but can perform the duties of either of the other two. For example, if you select the center console as the unit that depicts the digital video presentation of the radar information, the console on the left can be used to display electronic flight progress strips. The one on the right could then depict weather information or be used to select an exploded view of a small area. These exploded views operate similar to the "windows" in a word processing system in that they provide what one manufacturer describes as "viewports" of specific areas on the scope. This could give the controller the ability to watch a particularly critical operation more closely as part of the normal scan, or it could be used to monitor the activity on the final approach course of the destination airport to which he or she is vectoring.

Above each console is another video display terminal on which several other options can be selected. Similar to our present information display system (IDS), this unit can depict various pieces of information that are stored in memory. Some facilities use it to depict the approach profiles of the airports within the controller's area of responsibility. Others display frequencies of navaids in the area, airspace charts for the area, or surveillance approach information for any runway in an emergency. One facility even has a list of the over-the-counter medicines that controllers are allowed to take when they have a cold.

Each of the component displays is controlled by a keyboard, touch pad, cursor, light pen, or some other form of interface with the computers. Depending on which design you might see, you might have any or all of these control devices. I heard one technical representative suggest that the day will come when we will be able to talk to our computers. (Voice interactive computers . . . finally I can get some of my questions answered.)

The equipment is capable of displaying the weather in various intensity levels. The multicolor displays will allow different functions to be displayed in colors chosen by the controller. In fact, almost all material that I have seen or read about for sector suite components emphasizes how the equipment is designed to adapt to the personal preferences of the controller. It also touts how different-sized people will be equally at ease with the units.

Each position may have equipment that will allow a controller to select the radio frequencies that will be used from that position, the telephone communication lines that will be needed there, and the locations from which they will be selected. These communication components are referred to by various design names, but they fall into two general categories. The first, *integrated communication switching system* (ICSS), is currently in place at a few locations around the country. It is a slightly older system, is being phased out, and will be replaced by the newer generation of communication systems. The second, which I am given to understand is the system of the future, is called *voice switching and control system* (VSCS). Both of these components are designed to give the controller the flexibility of choosing the frequencies and telephone lines that will be used at that position. This concept makes each control position able to become any sector in that facility's airspace plan. This is an immense improvement over what we have now. Trying to change sector/position locations in our current system requires a major rewiring project, when it can be done at all.

Sector suites are intended to use off-the-shelf hardware and software, be relatively maintenance free, allow controllers to work together using electronic communication (thus reducing the noise levels in a large facility), and be adaptable to new technology as it comes along—every controller's dream of what he needs

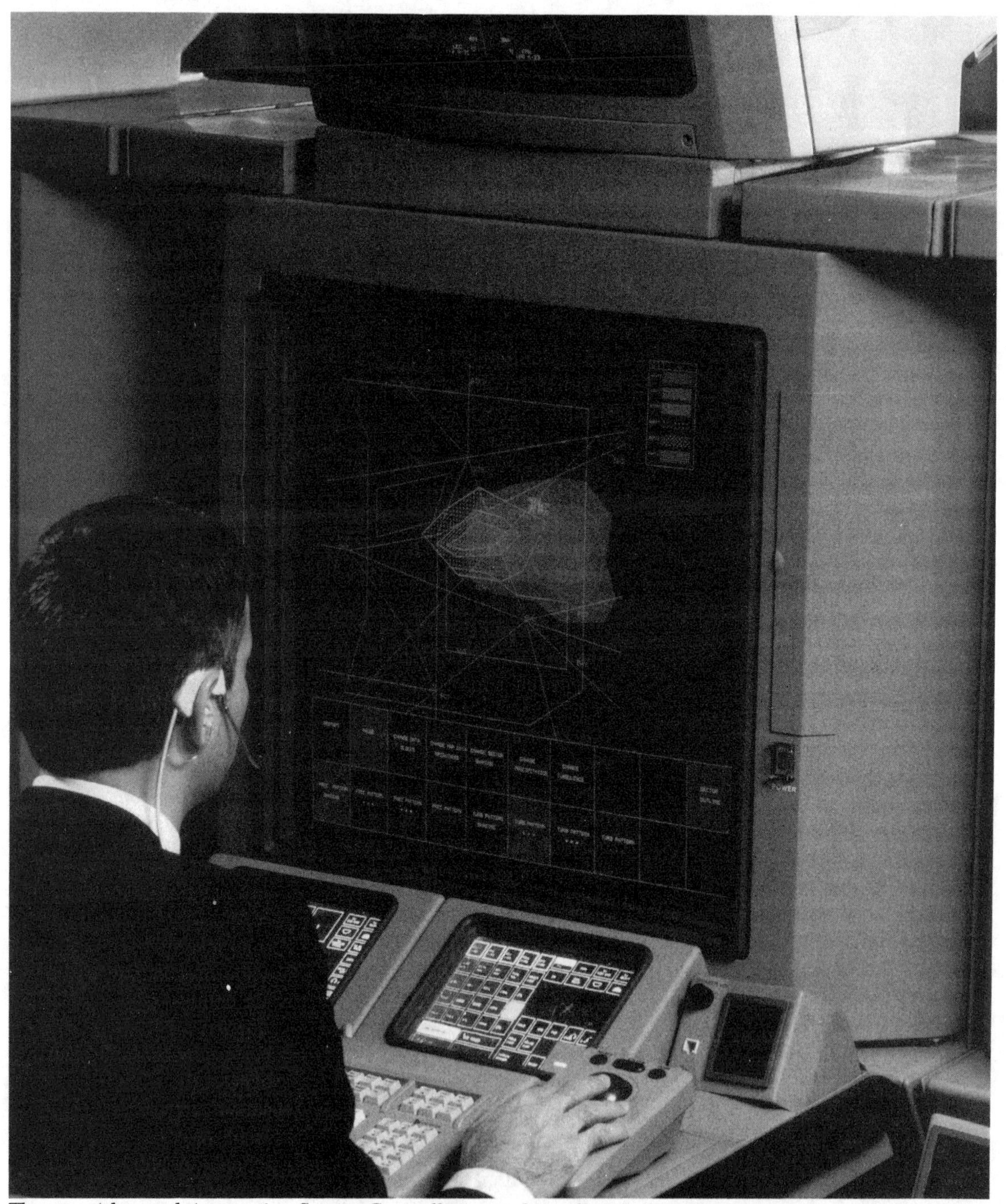

The new Advanced Automation System Controller consoles will include a high-resolution, 20-in. × 20-in. monitor with 2000 × 2000 picture elements displayed. By contrast the finest quality commercial TV set available today displays 480 × 400 elements. The console shelf will be designed to accommodate different heights of controllers, and the consoles will be used in both enroute and terminal radar facilities.

at a position of operation. (Hopefully they will also be designed to exist in the real controller's world. It would spoil my whole day if a spilled cup of coffee or a random lightning strike turns that dream into a nightmare.)

CONSOLIDATION

While all of this is impressive, you might be wondering why such a sophisticated system is needed if we just need to upgrade the units that the controller is using. The answer is that not only is the equipment changing, the concept of a control position is changing as well. Currently, if you control traffic in the Atlanta area, you work in Atlanta Tower or Atlanta Center. If you control traffic in Macon or Columbus (Georgia) or any number of other cities, you work in those respective cities. There are two reasons why the system works this way.

First, that is simply the way that the system has evolved over the years. Individual control facilities were established as the need for them came into being. In many cases, there was simply no way to determine where the next population and air traffic boom was going to take place. At that time, there were only a few locations in the country where there was enough concentration of people and aircraft to make the construction of a combined facility feasible. When possible, the FAA established common traffic facilities in locations where the radar coverage was sufficient to serve several locations. The New York Common IFR facility was a good example of this.

Second, the technology to combine the more widely spaced facilities either did not exist or was not adequate to the task when the individual facilities began their operations. It was more cost effective in these locations to train the controllers to serve the multiple jobs of tower cab controller and radar controller. This is no longer the case.

The technology now exists to locate the radar approach control function of several nearby terminals at one central site, and the agency is beginning to move in that direction. There will, of course, still be a need to staff the control towers, but the fact of the matter is that during a typical day, only a few controllers are actually needed in the tower cabs in those cities. The majority of the other controllers who work that traffic never leave the radar room, and they could be working those aircraft from a control position hundreds of miles away from that airport—enter the concept of *area control facility* (ACF).

The FAA will either build a mini traffic control center in a central location around several moderately sized terminal facilities, or choose selected major terminal sites to house the radar functions for several smaller airports in their immediate vicinity.

In the first case the concept, as I understand it, will be to expand the existing ARTCC buildings to accommodate the central sites. These will either be incorporated into the overall sectorization design of the center, or specific approach control sectors could be established which will operate much the same way that TRACONs work today.

In the latter case, a place like the existing Atlanta TRACON would be expanded to incorporate positions that would control the traffic at Macon or Columbus in much the same way that the controllers at these facilities do now.

In either case, the people who work the aircraft into several different airports will live and work in one geographic area. This would create several situations which, I think, will have major benefits in the long run. First, the cost of operating numerous facilities has an adverse impact on the total amount of money that is available to improve the equipment in use in the system. This type of operation would centralize much of the maintenance requirements at one location. Even if the number of technicians required to maintain the equipment does not change, the support staff required to operate all of those different sites could easily be reduced. It would require fewer buildings, vehicles, and personnel to perform the same functions, and the cost savings would be enormous. This savings will make more money available for improvement—assuming, of course, that the government doesn't reduce our budget because we don't have as much to maintain.

Lest you be concerned about the people aspect of this move, I think, and I hope the agencies will agree, that all of the positions vacated in this combination could be covered by attrition over the amount of time that it would take to accomplish this task. This consolidation would also greatly reduce the number of times that an individual would have to move his/her domicile to further career aspirations. There would, of course, be an initial major consolidation movement, but after that takes place, the savings in paying for people to move about would be in the millions of dollars each year. The savings for the people making the move would be equally as impressive. When I think back to the 7 percent mortgage that I had on the first house I bought after joining this agency, the subsequent moves all over the country, and the cost of things today, I just wish.

Bonus Benefits

Consolidation could also have many less-tangible benefits for pilots and controllers. (Let me again remind you that this is strictly my opinion and is not a part of any planned progression on the part of the FAA to which I am privy. It is just how I see this developing.)

There are three phases in a typical air traffic controller's career. (I am talking now about the person who is a controller and would like to remain a controller. Others among us start out as controllers with visions of other things in the future. But there is a group of controllers for whom controlling traffic is the main goal and they do it for the sheer pleasure of that job.) These individuals normally begin as young, hot-shot controllers who have the reflexes of cats, can visualize things with remarkable clarity, and can move airplanes on pure ability. They seem to inherently know ATC and have eyes in the backs of their heads.

Later, these individuals develop some experience that offsets a slight reduction in their reflexes, use techniques that have been learned over the years to maintain an impressive skill level, and develop an ability to anticipate rather than react.

Finally, as the skills begin to erode with age, they either retire from active air traffic control and do something else, or begin to look for other ways to continue their contribution to the job that most of them really like to do.

Don't take this totally out of context and think that all older controllers can no longer do the job, because this is not the case. Some of the people that I personally consider the best controllers in Atlanta, and the ones that I want on the difficult positions when things get rough, are a lot closer to 50 than they are to 30.

As for those controllers who have to make a decision, some become supervisors, managers, or staff specialists—hence the adage that, "Old controllers never die, we just take a lower profile." While these individuals remain close to the job and a few realize that this is what they really wanted, some of them miss the actual air traffic control and would rather welcome a chance to remain in that job at a slightly slower pace. It is rather unfortunate that by the time they reach the age where the skills start to erode, they have often been promoted to the place where they are most in demand.

The establishment of ACF locations will effectively establish a large number of VFR control towers at locations where busy TRACON facilities had once existed. These control towers are where I see the future potential of the FAA being developed. A lot of the terminal controllers that we hire in the future will initially be assigned to these towers. I think these facilities can become a place where the experience of the older controllers could be allowed to blend in with the new blood and be used to pass on some valuable experience. These places will not have the pressure of a busy radar function and will be ideal locations to allow experienced controllers to extend their usefulness in the air traffic control profession. If the FAA will allow some senior controllers to move to these locations without a significant loss of pay, we can develop a core of capable people who are where they want to be at that point in their career. These people will also have a more positive attitude because they will not be stuck where they do not want to be. We can then permit rookie controllers a chance to learn the basic skills that form the framework of well rounded air traffic controllers from the experts who already are. This is what I consider good human resource management.

I mentioned earlier that this consolidation could have a benefit to be reaped by pilots as well, and by now you are wondering how this is going to come about.

First, seasoned veterans are not going to make the mistakes that new controllers will make. They won't miss the takeoff slots or blow the sequence as often. They know how to make minor adjustments that smooth out the operation, and their experience allows them to have alternative plans which expedite operations. They are also more inclined to become acquainted with the people they work with and more involved in community and airport affairs. This familiarity means that they will take a personal interest in seeing that the best job is being done because they will be working with people they know. Those pilots who base their operations at a particular airport will quickly find themselves becoming accustomed to the style of these individuals, and there will be a natural stability at these locations.

Second, the new, young controllers will provide a breath of fresh air with their exuberance and vitality. You will quickly find them taking on the characteristics of the veteran controller, and then you will have the best of both worlds. They will have the advantage of learning the tricks from people who have prolonged their careers by knowing all the tricks. This will provide young controllers the ability to combine natural talent with experience and wisdom that would otherwise take years to develop, and we will have better controllers as a result. Also, with the prospects of advancement being a one-time move, these new controllers will develop the kinds of habits that we, as supervisors, like them to have, long before they develop the cynicism that seems to come from having to bounce around all over the country.

Well, enough of my personal philosophy, let's discuss a few more tomorrow's toys before we close out this book.

ADVANCED AUTOMATION SYSTEM

The collective label for all of these new types of equipment has come to be known as the *advanced automation system* (AAS). This equipment, or more precisely, this group of equipment components, will bring all of the information from most of the new devices together in a format that makes it useful to the air traffic control specialist. The system is made up of new computers, digital data-link communications, information processing devices, and all sorts of other components that will make several things possible.

First, controllers will have instant access to much of the data that we have been discussing so far. They will be able to call up things like multiple-site weather information. There will be virtual real-time flight plan processing and instant communications capability across a full range of frequencies and/or land lines. Additionally, realistic projections of flight path conflicts for all tracked targets will become a reality. What we use now is the best that you will find in the world, but the potential of this new equipment will make it look like stone age tools by comparison.

Second, the ability to extract, analyze, and project the data in this system will result in the ability to conduct realistic air traffic control across the entire country at the same time. Currently, our flow controllers can see the system as it is and factor the known scheduled traffic and the actual weather into their traffic management planning. By this I mean that they can look at the position of any aircraft that is airborne in the system and see when that aircraft will impact the traffic picture at the destination. They also know the departure times and scheduled arrival times of aircraft that fly those routes regularly. They currently use this information to meter the traffic flow by issuing a ground delay to an aircraft whose arrival time will overload the acceptance rate of the destination airport. The new departure time is calculated to generate an arrival time compatible to when the aircraft can be accommodated.

While this system does work, it lacks the ability to adjust to rapid changes in the system caused by unforecast weather or other unforeseen factors that impact an airport's acceptance rate. The new AAS will permit *user preferred trajectories* or more simply, an increased ability to go direct from point A to point B. The system will be able to detect conflicts in these paths far enough ahead for the controller to have time to react. As a result, it will significantly change the enroute time required for an aircraft to travel from origin to destination. For example, rules will automatically change every time new thunderstorms cut off existing routes. Or departure clearances may be delayed or accelerated based on unfavorable/favorable winds aloft and weather. The AAS will rapidly adapt to these events and, when delays are necessary, will plan those delays at high altitudes where fuel is conserved. In the end, this kind of flow management planning should minimize overall delays.

Beyond the equipment itself, AAS has become almost a philosophy. New equipment is analyzed to determine its ability to support AAS, and designs are changed when that determination is negative. The entire concept of air traffic control is moving in the direction of interactive, ''artificially intelligent'' components to not only support today's needs, but be able to adapt to changes that will occur tomorrow. There will come a time when a controller will be able to ask a computer any intelligent ATC-related question and be given an answer and a series of options.

OTHER DEVELOPMENTS

A lot of developments are also occurring in areas that we do not normally associate directly with aviation. For example, new technology has been developed using vapor detectors to screen carry-on baggage for flammable liquids, and a process called *thermal neutron activation* (don't ask me for a technical explanation) has been tested for use in checking baggage for explosives. There are also studies in the field of aviation medicine designed to reduce the effects of stress and fatigue on crewmembers. Other work is being done which should improve the accident survivability. In line with the survivability factor, there is renewed interest in developing a non-misting fuel for aircraft, or at least one that is less subject to explosive burning during a crash. This wide diversity of study is just the tip of the iceberg in terms of the things with which the FAA is involved. They are studying things like propulsion technology, transatmospheric vehicles capable of low-orbit flight, advanced metal matrix or high-temperature carbon composites for structures, and they are even studying how to make better concrete for the runways on which you land.

Where does the average pilot fit into all of this change? Unfortunately, the typical pilot looks at this and says, "Sure, you guys are building all of this hi-tech expensive junk that little people like us are going to have to put in our birds if we ever want to fly again." You are partially right. You probably will have to have some new equipment, but consider the following.

In the 1970s I bought a brand new piece of equipment that had just come on the market which changed forever the way engineers and accountants would do their jobs. This device was about the size of a large wallet, was extremely fast at calculating simple arithmetic functions and cost almost $200.00. Today you can buy a multiple function solar calculator in any department store for about $15.00. These newer units are a fraction of the size of the first ones and are infinitely more sophisticated.

This same scenario is taking place in the aviation market. The first *area navigation* (RNAV) systems were bulky, had limited capabilities, and cost thousands of dollars. Today you can purchase a *long-range navigation* (LORAN) system that is programmable for several dozen locations, accurate to a tenth of a mile, and can be purchased for only a few hundred dollars.

Shortly, the typical pilot will be able to put together an equipment package that includes MLS/ILS equipment, a relatively low-cost LORAN system, a 360-channel radio, and a Mode S transponder for an affordable price. Put this in a single-engine aircraft with long-range fuel tanks and you will have an aircraft that has some of the sophistication of the big airliners, will go almost anywhere, and will still be within the budget of the average pilot.

You don't need an expensive aircraft in which to install this equipment either. One of the controllers who works with me has a Cessna 140 with an instrumentation package that contains a good radio, a Mode C transponder, and a LORAN. He takes this aircraft all over the country, and I believe he would agree that he is what you might call a typical pilot. I think that he is indicative of the type of individual who is not afraid of the future of aviation and knows how to combine the beauty of an old aircraft with the technology of tomorrow's requirements.

Final Thoughts

I HOPE THAT THIS BOOK HAS HELPED YOU BECOME AWARE OF SOME OF THE ATC capabilities that exist for your use. Controllers take considerable pride in the fact that we are able to respond to the needs of pilots. Regardless of how much we grouse and complain about pilots, you will always get the best possible service that we are able to provide, given the traffic load and our ability to do so.

Earlier I told you that there were three types of errors. Now let me add a fourth. Our society gives us ample opportunity to be guilty of the worst possible kind of sin—doing nothing in the face of evidence that we should be doing something—the ''bury your head in the sand and hope it gets better'' approach. Pull your head out of the sand, if that's where it is, and get out there and become a positive force for change in our industry.

I would like to leave you with one thought that might make you a little more comfortable working in the ATC system. Even though we air traffic controllers collectively work millions of aircraft every year, we have never yet left one up in the air.

Thanks, and hopefully we will talk again.

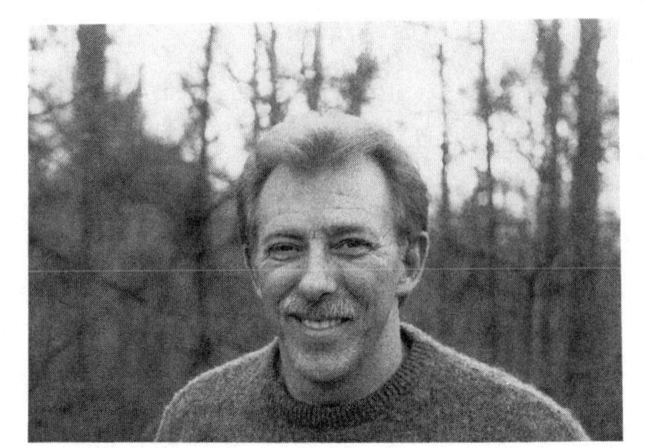

About the Author

JOHN STEWART IS CURRENTLY A SUPERVISORY AIR TRAFFIC CONTROLLER AT the William B. Hartsfield Airport in Atlanta, Georgia, and resides in Douglasville, Georgia, with his wife Peggy and their four sons. In addition to his current assignment, Mr. Stewart has held positions as a controller and procedures specialist at Atlanta; a supervisor at Standiford Field in Louisville, Kentucky; a controller and training specialist at Raleigh/Durham Airport in Raleigh, North Carolina; and as a controller at Bishop Airport, Flint, Michigan. He has authored several publications for the FAA including the Training Handbook and Procedures Handbook used respectively in Raleigh and Atlanta. Mr. Stewart spent five years in the U.S. Air Force, holds a degree in Science from C.S. Mott College in Flint, Michigan, and began his civilian aviation career as a commercial pilot and flight instructor for Welch Aviation in Alpena, Michigan. He helped found, and then served as chief flight instructor for Shephard Aviation in Midland, Michigan, and taught college-level aviation courses in community colleges in Alpena and Midland, Michigan. Mr. Stewart has accumulated 2500 hours of flight time in various categories of aircraft in over 20 years of aviation experience and has conducted or participated in numerous pilot/controller forums and ''Operation Raincheck'' seminars during that time.

Index

A

abbreviations, ix-xi
active radar returns, 80
additional services, 51-55
advance filing, 6-7
advanced automation system (AAS), 216-217
affirmative, definition of, 21
air route traffic control center (ARTCC), 6
air traffic control, 29-30, 146-147
Air Traffic Control Handbook, 127
airborne evaluations, 124
aircraft identification, 15, 21-23, 81, 107
airfiling, 29-33, 145-146
Airman's Information Manual (AIM), 20, 110, 127
airport advisory area, 177-178
Airport Facility Directory, 38, 96, 110, 167, 170, 173
airport radar service area (ARSA), 172, 187-188
airport surface detection equipment (ASDE), 202-203
airport surveillance radar (ASR), 71
airport traffic area (ATA), 55, 163
airspace, 157-190
  airport advisory area, 177-178
  Airport Facility Directory listing for, 167
  airport radar service area (ARSA) and, 172, 187-188
  alert areas, 178
  anatomy of, 164-165
  approaches, 168
  ATC, 29-30
  chart designations for, 166-167
  common sense use of, 188-190
  control zones, 175-177
  controlled, 138
  controlled firing areas, 178
  designation of, 29-30
  division of, 10
  elements of, 163
  FSS and, 170
  ground controlled approach (GCA), 29
  military operations areas (MOAs), 173, 178-181
  minimum safe altitude listing, 174
  radar map showing approaches, 185
  regulations concerning, 161-164
  restricted areas, 178
  special use, 138, 178
  terminal control area (TCA) and, 172, 181-183
  terminal radar service areas (TRSA) and, 171, 187
  towers and, 169-170
  transition area, 168

Victor airway, 168
visibility and cloud clearances in, 159-161
warning areas, 178
alert areas, 178
altitude, 83, 108, 121, 127, 129-130
altitude deviation information, 52
ambiguous communications, 42
approach control, 6, 94, 170
approach gate, 148
approach wall, 91
approaches, airspace designations for, 168
area control facility (ACF), 213, 215
area navigation systems (RNAV), 218
ARSR-4 radar, 199
ARTCC, 13, 170
ASR-9 radar, 199
ATC equipment, 69-118
facility layout of, 90-97
Flight Data Systems, 104-118
pilot awareness of, 85
radar as, 70-85
radio and telephone systems, 85-104
tracking equipment, 81-85
ATC instructions vs. pilot authority, 142-145
attenuation, 78-80
authorized deviations, 135
automated radar terminal system (ARTS), 7, 14, 81-85
automated weather observing system (AWOS), 209
automatic dependent surveillance (ADS), 202
automatic terminal information service (ATIS), 12, 34-36
azimuth, radar, 70

B

background video circuits, 78
base, TCA, 162
basic pilot errors, 1-16
beacon control slash, 80
beacon replies, 80
bird activity information, 52
braking action, 155

C

call signs, 20-23, 81, 107
certified flight instructors, 2-4
chaff information, 52
charts
airspace designations on, 166
approaches shown on, 168
ARSA on, 172
control tower on, 169-170
control zone on, 171
frequency designation on, 96
FSS on, 171
military operations areas (MOAs) on, 173
military training routes (MTR) designation, 180
minimum safe altitude listing, 174
NDB, 168
terminal control area (TCA) on, 172
TRACON on, 171
TRSA and, 171-172
TVOR, 168
circular polarization, 77
civil aircraft sonic boom (FAR 91.55), 134
clearances, 11, 55-56, 142, 145
cruise, 62-63
special VFR, 58-62
taxi, 12
cloud clearance, 159-161
clutter, radar, 75
collision avoidance systems (TCAS), 205
commercial telephone, 102
commission errors, 1
common digitizer, 71
common traffic advisory frequency (CTAF), 169, 178
communications, 17
acknowledging reception of, 18-20
ambiguous, 42
concentrating on, 23-24
ignoring or misinterpreting, 56-58
improvement of, 6-9
nonverbal, 27-28
proper use of, 38-39
proper word usage in, 17-18
roles of controller and pilot in, 25-27
standard procedure vs. individual requirements, 24-25
understanding instructions, 18
using and abusing system, 34-38
computer identification number, 107
consolidating new technology, 213-216
contact approach, 149-150
continental control area, 163
contract employees, FAA, 198-199
control tower, 93, 169-170
control zone, 163, 171, 175-177
controlled airspace, 138, 159-161
controlled firing areas, 178
controllers
errors and, 2
improving communications with, 6-9
inexperienced or student pilots and, 5-6
pilot requests and, 9-14
review process for, 124-125
role of, 66-67
coordinated universal time (UTC), 12, 107, 110
coordination fix, 109

course cards, 204
cruise clearance, 62-63

D
data control unit (DCU), 32
delays, 63-66, 112, 152
departure control, 94, 107, 109, 110
heading designation for, 12
intersection departures, 150-153
designated floor, TCA, 162
destination airports, 112
dial circuitry, 99
digitized radar, 71
direct altitude and identity readout (DAIR), 81
discrete frequency, 86, 87
distance measuring equipment (DME), 76
Doppler radar, 77, 208
duplicate flight plans, 115-118

E
emergency frequencies, 103
emergency patches, 83
emergency, definition of, 47-51
enroute flight advisory service (EFAS), 38
enroute navigation, 13, 30, 83
equipment limitations, 53
equipment requirements, TCA, 141
equipment usage, 11
estimated departure clearance times (EDCTs), 112-115
evaluation branch, FAA, 123-124

F
FAA Aviation News, 188
FARs, 120, 128-129
locating references to, 128
recommended vs. mandatory, 132
Federal airways, 163
Federal Aviation Administration, 2, 123-124
contract employees for, 198-199
future developments for, 192-196
procurement procedures for, 193-196
re-structuring of, 196-198
Federal Register, 139
fixes, 109
flight data encoding printer (FDEP), 32, 104
flight data input-output (FDIO), 32, 92, 104
flight data systems, 104-118
flight plans
advance filing of, 6-7
airfiling of, 29-33
allowing adequate time for entry of, 33-34
ARTS IFR activation of, 14
cancellation of, 12
computer entry of, 13, 28, 31-33
duplicated, 115-118
errors in, 28-34
FAR 91.83 regulations and procedures, 145
filing, 11, 28, 30, 33, 121
IFR, 145
incorrect data in, 28
regulations and procedures and, 127
truncating, 111
UTC in, 12
flight progress strips, 104-112
flight service stations (FSS), 9, 87, 170-171
flight standards district office (FSDO), 10, 126
flight training, 3, 4
Flight Watch, 38
frequency
assignment of, 86-87, 96
chart designation and, 96
discrete, 86, 87
emergency, 103
FSS, 87
joint-use, 89
management of, 86-88
military, 89-90
TCA, 141
UHF, 88
UNICOM, 87
VHF, 88
VHF emergency (121.5), 87
VOR, 9
future outlook, 191-218

G
general aviation district office (GADO), 61
general operating and flight rules (FAR Part 91), 128
geosynchronous satellites, 204
go ahead, definition of, 45
ground controlled approach (GCA), 29, 73
ground school, 3

H
handing off, 9, 13
headings, 155
holding pattern information, 52
holding speed, 133
hot lines, 100, 101
human error, 1

I
IFR, 7, 83
airfiling, regulations and procedures for, 30-33, 145-146
practicing, 6
ignorance errors, 1
immediate, definition of, 46-48
inbound strips, 106
information display system (IDS), 92, 194
information transfer procedures, 61

instructions, 142
instrument flight rules (FAR 91-B), 129
instrument landing system (ILS), 18
instrument military training routes, 173
integrated communication switching system (ICSS), 211
interception final, 148-149
International Civil Aviation Organization (ICAO), 173
intersection departures, 150-153
IR routes, 179

J

jargon, 41
jet routes, 163
joint-use frequencies, 89

K

key packs, 86, 88-89, 92
  telco, 97-102

L

large aircraft
  intersection departure and, 152
  TCA and, 140
Letter to Airmen (LTA), 25
light signals, 27, 167
linear polarization, 77
locator outer marker (LOM), 18
logarithmic fast time constant (LOG FTC), 78
long-range navigation (LORAN), 218

M

mandatory regulations and procedures, 132
master positions, 90
McIntyre, Harry, 25
microwave landing systems (MLS), 205-208
military frequencies, 89-90
military operations areas (MOAs), 173, 178-181
military training routes (MTR), 179, 180
minimum vectoring altitude, 149
minimum VFR visibility and distance from clouds, 158
Mode C, 52, 83
monitor position circuits, 101
multiplexing radar, 74

N

National Airspace System, 163, 164
National Oceanic and Atmospheric Administration (NOAA), 166
negative, definition of, 21
next-generation weather radar (NEXRAD), 208
night flights, 27
no procedure turn (NoPT), 57
nondirectional beacon (NDB) approach, 149, 168
nonstandard phraseology, 41
nonverbal communications, 27-28
NOTAMs, 36-38

O

oceanic display and planning system (ODAPS), 204
omission errors, 1
Other Airspace Areas, 177, 179
overflight strips, 106
override circuitry, 98-99

P

passive radar reflections, 80
patches, 83, 84, 85
phraseology, 20-21, 41-67
pilot authority, ATC instructions vs., 142-145
pilot errors, 1-16
pilot monitoring, 126-127
pilot requests, 9-14
pilot requirements, TCA, 140
planned ground delay programs, 63
planning, 11
polarization, circular vs. linear, 77
position reporting, 122, 130-13
positive control area, 163
precision approach radar (PAR), 71, 73
preferential departure route (PDR), 110
preflight, 4-5, 11
proper word usage, 17-18, 41-67
proposed outbound strips, 106
push-to-ring circuitry, 100

R

radar, 13, 70-85
  active (secondary) returns, 80
  active vs. passive detection in, 70
  advances in technology for, 199-202
  airport surface detection equipment (ASDE), 202-203
  airport surveillance (ASR), 71
  alphanumeric displays for, 81
  ARSR-4, 199
  ASR-9, 199
  attenuation, 78-80
  azimuth in, 70
  circular polarization of, 77
  clutter and, 75
  digitized, 71
  display of, 82
  DME and, 76
  Doppler, 77, 208
  LOG FTC, 78
  loss of, 53
  map showing, 72, 185
  moving target indicator, 75-77

multiplexing, 74
next-generation weather (NEXRAD), 208
operation of, 70
passive reflections, 80
precision approach (PAR), 71, 73
processed information and, 80
rain and water and, 76, 77
range of, 70
real-time information and, 80
scopes for, 73-75, 200-202
secondary radar beacon target (SECRA) in, 71
shadows, 74
standard design of, 86
symbols on map of, 72
terminal Doppler weather (TDWR), 208
thunderstorms and, 79
TPX-42, 81
transmitter-receiver for, 86
transponders and, 70
video map overlays and, 80
wave travel and path of, 75
weather effects and, 77-78
white out and, 77
radar beacon target, 71
radar contact, definition of, 44-45, 52
radar mile, 70
radio and telephone systems, 85-104
radio position, 97
radio systems, 86-97
key packs in, 88-89
remote transmitter-receiver for, 89
rain, radar and, 76, 77
range, 70
range gate, 75
real-time information, radar and, 80
recommended regulations and procedures, 132
reduced intercept interval, 148
refresher courses, 3
regulations and procedures, 119-156
altitudes and, 129-130
ATC instructions vs. pilot authority, 142-145
ATC procedures, 146-147
contact approach, 149-150
controller participation in upgrading, 124
FARs, 120
following procedures when requesting special procedur, 154
IFR airfiling, 145-146
intercepting final, 148-149
intersection departures, 150-153
pilot monitoring, 126-127
position reporting, 130-133
quality control in, 123-127
recommended vs. mandatory, 132
request for reductions in, 147, 152
requesting authorization for illegal procedures, 153
severe weather and, 144
speed limits, 133-138
staying current with, 122-123
technical violations of, 143-145
working for changes in, 125-126
reliever airports, 113
restricted areas, 178
ring-down circuitry, 100
roger, definition of, 20
runway length, 167

S

satellites, 202-204
scopes, radar, 73-75, 200-202
scratch pad data, 83
secondary radar beacon target (SECRA), 71
sector suites, 210-213
security technology, 217
sequencing procedures (see delays)
service volume, 168
severe weather
departure times and, 112
regulations and procedures and, 144
shadows, radar, 74
signals, 27
single-frequency approach (SFA), 90
sonic booms, 134
spacing procedures, 63-66
special use airspace, 138, 178
special VFR
airport advisory area and, 177
clearance for, 58-62
speed limits, 133-138
authorized deviation from, 135
under TCA, 137
squall lines, radar and, 78
standard instrument departures (SIDs), 24
standard procedures, individual requirements vs., 24-25
standard terminal arrival routes (STARs), 24
student pilots, controllers and, 5-6
syntopicon, 128
systems deviation, 33
systems error, 33

T

tab list, 33
tangency, 75
technical violations, 143-145
telco key packs, 97-102
telephone systems, 97-104
improvisation with, 102-104
land lines, 103

terminal control area (TCA), 9, 138-142
  airspace of, 162
  base and designated floor of, 162
  equipment requirements for, 141
  FAR 71 definition of, 138
  frequencies for, 141
  future developments for, 205
  inner ring of, 140
  large, turbine-powered craft, 140
  operating rules for, 139-142
  pilot requirements for, 140
  remaining clear of, 181-183
  specifications for, 139
  speed in and under, 137
  structure of, 181-187
  transition to, 172-173
  transponder requirements, 141
  unauthorized penetration of, 10, 181-183
  within vs. under, 140
terminal doppler weather radar (TDWR), 208
terminal radar approach control (TRACON), 171
terminal radar service areas (TRSA), 171, 187
terminal VOR (TVOR), 168
thermal neutron activation, 217
thunderstorms, radar and, 79
TPX-42, 81
tracking systems, ARTS, 81-85
TRACON, 214, 215
traffic advisories, 52
traffic alert and collision avoidance system (TCAS), 205
transcribed weather broadcasts (TWEB), 38
transfer of control-communication fix, 109
transition area, 162, 163, 168
transponders, 7, 80
  codes for, 13, 70, 107
  TCA requirements, 141
truncated flight plans, 111

U

UHF frequencies, 88
uncontrolled airspace, 163
  visibility and cloud clearance in, 159-161
UNICOM, 87, 166, 167, 169
updating facilities, 192-196
urgency, definition of, 46-48
user preferred trajectories, 217

V

vectors, 9, 10, 52, 53, 54
VFR, 7, 83
  airport advisory area and, 177
  flight progress strip, 108
  minimum visibility and distance from clouds, 158
  practicing, 6
  reporting points, 8
  weather minimums for, 161
VHF emergency frequency (121.5), 87
VHF frequencies, 88
Victor airway, 168
video map overlays, 80
visibility, 159-161
visual military training routes, 173, 179
voice switching and control system (VSCS), 211
voice-call circuits, 100, 101
VOR, frequency for, 9

W

warning areas, 178
weather, 52, 53, 77-80
weather minimums, VFR, 161
weather technology, 208-210
white out, definition of, 77, 78
wilco, definition of, 21
winds aloft, 121
word concepts, 41-67
written exams, 3